GW01066266

The Limited

The Story of the Cornish Riviera Express

'Steam Past' Books from Allen & Unwin

THE LIMITED by O. S. Nock
THE BIRTH OF BRITISH RAIL by Michael R. Bonavia
STEAM'S INDIAN SUMMER by George Heiron & Eric Treacy
GRAVEYARD OF STEAM by Brian Handley
PRESERVED STEAM IN BRITAIN by Patrick B. Whitehouse
TRAVELLING BY TRAIN IN THE EDWARDIAN AGE by Philip Unwin
MEN OF THE GREAT WESTERN by Peter Grafton

The Limited

The Story of the Cornish Riviera Express

O. S. Nock, B.Sc., C.Eng., F.I.C.E., F.I.Mech.E.

London
GEORGE ALLEN & UNWIN
Boston Sydney

First published in 1979

GEORGE ALLEN & UNWIN LTD
40 Museum Street, London WC1A 1LU

British Library Cataloguing in Publication Data

Nock, Oswald Stevens
The Limited.
1. Cornish Riviera Express
I. Title
385'.09423 HE3019.W/ 78-40572

ISBN 0-04-385073-1

Picture research by Mike Esau

Book designed by Design Matters

Typeset in 11 on 12 point Imprint by Bedford Typesetters Ltd
and printed in Great Britain
by W & J Mackay Limited, Chatham

Contents

Illustrations

Preface

To write the story of the 'Cornish Riviera Express' is like recalling the activities of a lifelong friend. I first saw it on the outskirts of Reading when I was a child in a push-chair, and in the years 1913–14 I used to see the up train come through Reading West while I was waiting for the motor to take me home from prep school to Mortimer. In later years, as one of its passengers, I did some of my earliest stop-watching and in more recent times I rode its locomotives and witnessed some exciting trials from the privileged position of the dynamometer car.

It was in its many-sided importance that the 'Limited' had such an outstanding appeal: as holiday train *par excellence*; as prestige symbol in the technical aspects of operation; as a supreme challenge to the locomotive department. Its character was largely unchanged for upwards of fifty years; but today, if it has not yet come to partake of the glamour of present-day railway working in the form of 100 mph electric locomotives or HST rolling stock, its traditions may yet form the centre-piece of a great modernisation of travel to the West Country. Unseen, romance still brings up the eleven-thirty from Paddington.

Silver Cedars,
High Bannerdown,
Batheaston,
Bath.

O. S. NOCK
August 1977

I

Inauguration in 1904

At the beginning of the twentieth century Cornwall was to most people a remote, almost unknown land. The Cornish themselves lived in isolation and referred to the occasional tourists as visitors from England. But a phenomenal change was soon to come. The management of the Great Western Railway, rid of the incubus of the broad gauge, began literally to create a vast holiday traffic out of nothing. Until then Falmouth, already declining in importance as a port, was the goal of the Cornwall Railway; west of Truro, over the one-time West Cornwall Railway through the languishing tin-mining areas, the prospects of a flourishing holiday business were as dim as the look of the landscape. The magnificent coastal scenery of the Lizard and Land's End districts, lying far beyond railways, could be reached only by venturing over rough, hilly and alarmingly narrow roads. But the climate was genial and, although the possibilities of development were not those of the more obvious holiday resorts, there were more subtle charms; and when some genius at Paddington found that if the map of Italy was seen in mirror-image it looked extraordinarily like Cornwall the name 'Cornish Riviera' was born.

From that time onwards the 'Riviera' concept was plugged for all it was worth. It was a time when the prestige value of long non-stop runs was in the ascendant, but so far as the West of England services of the Great Western were concerned there was a good deal more to it than mere prestige. Through the many vicissitudes of the broad gauge era the 'Flying Dutchman' leaving Paddington at 11.45 a.m. had been the crack train to the west. But even before the epoch-marking year of 1892 an earlier express, leaving Paddington at 10.15 a.m., had been put on; it was named the 'Cornishman'. Before Plymouth it called only at Swindon, Bristol and Exeter, and when the refreshment rooms at Swindon were taken over, and a non-stop run from Paddington to Bristol made possible, the departure was made at 10.35 a.m. Penzance was reached at 7 p.m.

By 1896 the train had become so popular as to require running in two sections during the summer months and it was in 1897 that the advance portion, leaving Paddington at 10.30 a.m., was run non-stop to Exeter, a distance of 193·9 miles, in 3 hours 43 minutes – then the longest non-stop run in the world. Beyond Exeter stops were made at Plymouth, Lostwithiel (for Fowey), Par (for Newquay), St Austell and Truro, and the train terminated at Falmouth, completing the journey in a time

1. Coming events cast their shadow: the broad gauge North Mail hauled by the Gooch 4–2–2 engine, 'Warlock', in evening sunlight near Exminster, running over track newly relaid with transverse sleepers, replacing Brunel's 'baulk road'.

Locomotive Publishing Co.

62 minutes quicker than the broad gauge 'Flying Dutchman'. Both the 10.30 a.m. summer relief and the 10.35 regular 'Cornishman' were made up of corridor stock throughout, but it was not until 1899 that a restaurant car was included, and then only on the 10.30 a.m. These trains were generally hauled by the Dean 7 ft 8 in. 4–2–2 locomotives, though by the turn of the century the Atbara class 4–4–0s were entering traffic. With these trains and the corresponding eastbound services from Exeter to Paddington the Great Western were becoming practised runners of long non-stop journeys, and the great event of 1902 came perhaps as less of a surprise than it might otherwise have done.

The visit of King Edward VII and Queen Alexandra to the naval establishments at Dartmouth and Devonport was made a positively gala occasion by the GWR, and with the Royal train they made the longest non-stop runs that had ever been attempted: from Paddington to Kingswear, 228·5 miles, on 7 March 1902, and returning from Millbay Docks to Paddington, 246·7 miles, a few days later. The booked time between Paddington and Exeter was 209 minutes in each direction, 14 minutes faster than the advance section of the 'Cornishman' and an average speed of 55·7 mph. In both directions the train was worked by the Atbara class locomotive No. 3374, normally named 'Baden Powell' but

2. Start of the narrow gauge era in the west: one of William Dean's massive Duke of Cornwall class 4–4–0s, no. 3322, 'Mersey', that replaced the broad gauge 4–4–0 saddle tanks as maids of all work west of Newton Abbot.

Locomotive Publishing Co.

temporarily renamed 'Britannia' for the occasion. The speed was good, but it was no more than a mild curtain raiser to what was achieved rather more than a year later when the Prince and Princess of Wales, later King George V and Queen Mary, paid a visit to Cornwall and Royal saloons were attached to the first portion of the 'Cornishman', then leaving Paddington at 10.40 a.m. With one of the latest 4-4-0 locomotives, no. 3433, 'City of Bath', a very fast run was made non-stop to Plymouth, in 3 hours 53½ minutes for the 245·6 miles to North Road station. Exeter had been passed in the then record time of 172 minutes 34 seconds from Paddington.

This run, if nothing else were needed, established the outstanding quality of the City class engines, because in it, even more so than with the Royal train workings of 1902, something entirely new in Great Western locomotive operation was being attempted. It was not so much the length of run, though this of

course was a major factor in both coal and water consumption, but the nature of the route itself. Three distinct sections of the old Brunel broad-gauge network were involved: the Great Western proper, from Paddington to Bristol; the Bristol and Exeter; and the South Devon Railway. The first two, though including some stiff – though short – inclines, were laid out in the grand manner of Brunel with notably straight alignments and were level for the most part, or nearly so. The South Devon, on the other hand, was designed for the atmospheric system of traction, which it was believed could make light of any gradient. Knowing much of Brunel's character from erudite biographies and scholarly professional accounts of his work in many diverse fields, it is fascinating to speculate upon how different the South Devon line might have been had not the 'atmospheric' red herring not been drawn across the trail. Yet there is evidence that the 'atmospheric' was not the only factor that led to his choice of route. At one time there were proposals for a much straighter, better graded, and more direct line through the South Hams; but at an early stage it was discarded because of the high cost of some of the engineering works that would have been involved. The South Devon did not enjoy the exceptionally deep purse of the Great Western proper, and much of the work had to be done as cheaply as possible.

The abrupt change in the physical nature of the route that began immediately to the west of Newton Abbot gave rise to a pattern of locomotive operation that prevailed until that momentous year of 1904. Although the South Devon Railway began at Exeter, it was customary in broad gauge days for the 8 ft single-driver express locomotives that had brought the trains down from Bristol to continue round the level coastal stretch to Newton Abbot, and there to hand over to the ugly but effective 4–4–0 saddle tanks of the South Devon line, which had a monopoly of the traffic not only to Plymouth but in Cornwall as well. In the early years of the standard gauge new engine building followed in the same tradition with the massive 4–4–0 Duke of Cornwall class taking over from the Dean 4–2–2s, and next the 5 ft 8 in. Camels and Bulldogs taking over from the 6 ft 8 in. Atbaras. Equally, when G. J. Churchward, in 1901, produced his classic scheme for the complete modernisation of Great Western motive power, his famous diagram included a 6 ft 8 in. 4–6–0 with two outside cylinders for the ordinary heavy express traffic of the line, and an exactly corresponding design with 5 ft 8 in. coupled wheels for the line west of Newton Abbot. Because of its geographical situation, and the existence of the former South Devon Railway workshops there, Newton became one of the most important locomotive out-stations on the line, with a works capable of dealing with all except the heaviest repairs to locomotives.

The Royal train workings of 1902 and 1903 and the success with which the Atbara and City class engines coped with the exceptional gradients of the Dainton, Rattery and Hemerdon inclines opened up a new prospect in West of England train operation. With the moderate loads of the Royal specials, those 6 ft 8 in. 4–4–0s climbed the 1 in 40 and 1 in 50 gradients as competently as, elsewhere on the line, they ran freely at 75 mph on level track. In some respects there was in fact greater certainty. The long, level stretches of the Bristol and Exeter main line in the flat, almost treeless country of north-west Somerset are exposed to every wind under heaven, and a tearing westerly gale off the Severn Sea can be equal to two or three extra coaches on the train, whereas in the lush, sheltered combes of South

3. A blend of old and new: an up train crossing the Brunel timber viaduct across the beach at Penzance, running on 'baulk road' converted to narrow gauge, and hauled by the beautifully named Duke class 4–4–0 no. 3273, 'Armorel', about 1899.

Locomotive & General

4. Inauguration of the 'Limited' – 1 July 1904: the first down express passing Old Oak Common hauled by the French compound Atlantic no. 102, 'La France'. This engine was regularly used on the London–Plymouth non-stop run, turn and turn about with 4–4–0s of the City and Atbara classes.

Locomotive Publishing Co.

Devon it is only the gradients and the curves that have to be subjugated. The curves do indeed seem unbelievable to a visitor who has ridden down from Bristol on the footplate. One can readily appreciate also that Brunel adopted this extraordinary alignment to minimise the cost of cuttings and embankments. Furthermore, at the time he built the line, in the 1840s, the likelihood of its becoming a teeming artery of intense holiday traffic had not been considered, even in the far-sighted minds of those who pioneered the broad-gauge network.

The Royal special of July 1903 having shown clearly what could be done with Churchward's splendid City class locomotives, it became apparent that, in developing a holiday traffic to Cornwall, Plymouth would be a much more convenient engine changing point than Newton Abbot, quite apart from giving the Great Western the tremendous boost of a daily non-stop run in each direction of 245·6 miles. Nothing like it had previously been regularly scheduled anywhere in the world, and when it was definitely announced to be coming, in the summer of 1904, the management of the GWR felt it should have a special name, distinct from the existing 'Cornishman' and from the 'Flying Dutchman' and 'Zulu' of broad gauge days. A competition was announced in *The Railway Magazine* for a prize of three guineas, on which the adjudicator was James Inglis, the General Manager of the GWR. Provisionally it was known as the '3TF' – three towns flier: Plymouth, Stonehouse and Devonport. The competition was announced in the August 1904 issue, and the result announced a month later. A total of 1,286 entries were received, and on the short list were 'Cornish Riviera Limited', 'Cornubian', 'Lyonesse Limited', 'One and

5. Also on 1 July 1904, the first up 'Limited' leaving Penzance, hauled by the Bulldog class 4–4–0 no. 3450, 'Swansea'. Note the 'baulk road' is still in use over the viaduct twelve years after the final conversion of the track from broad gauge.
O. S. Nock Collection

All Limited', 'Royal Duchy Express' and 'Speedwell Express'; but Mr Inglis chose the 'Riviera Express', and the prize money was divided between two readers of *The Railway Magazine*. Actually there is no evidence of this prize-winning title being used officially, and from a very early stage in its career it became the 'Cornish Riviera Express' or, more familiarly to the railway staff, simply the 'Limited'; and at first its formation was very definitely limited.

From its inauguration on 1 July 1904, it consisted of six coaches, all of which went through from Paddington to Penzance. Except for the dining-car, which was one of the 30-ton elliptical roofed Dreadnoughts, the coaches were all of the Dean clerestory type, with the following make-up reading from the locomotive and tender: brake third; third; first and second compo; dining-car; first and second compo; brake third, with a total seating capacity for 24 firsts, 48 seconds and 128 thirds. The dining-car was available to all classes. The tare weight of the six coaches was 141½ tons, easily within the capacity of the City class engines east of Plymouth, and of the 5 ft 8 in. Bulldog class in Cornwall. The only intermediate stops were at Truro, Gwinear Road and St Erth, which caused a friend of mine to comment that the 'Limited' passed all the larger places, referring to St Austell, Camborne and Redruth, while calling at Gwinear Road. From the outset, however, the train was aimed at the tourist traffic, and Gwinear Road served Helston and the Lizard peninsula.

As originally run, the route provided a remarkable range of West Country landscapes, from the Thames valley and the Vale of the White Horse to the passing glimpses of a gracious Georgian city like Bath, and the intense railway interest in the slow perambulation of the Bristol avoiding line and the negotiation of such a maze of junctions. Then came the Somerset coastal flats with distant glimpses of the Mendip range and the distinctive cone of Glastonbury Tor, and so to Taunton, through the old Brunel station, and to the attack on the Blackdown ridge. Railway enthusiasts would know that it was here, in descending from the ridge, that the 'City of Truro' had attained a very high speed with the 'Ocean Mail' of 9 May that same year, though what that speed was had not then been fully revealed. In those early days the 'City of Truro' was not one of the engines regularly working on the 'Limited', because she was stationed at Exeter, and it was worked mainly by London engines and men. The first of the French compounds, No. 102, 'La France', was a favourite on the job, though the 'City of Bath' and several of the Atbaras took their turns.

One always felt that the threading of the crest of the Blackdown ridge in the Whiteball tunnel marked the true entry into the West Country, because apart from entering the tunnel in Somerset and emerging in Devon one became at once conscious of the red soil, the simpler, rather austere towers of the village churches, and the intense green of the vegetation. Exeter itself could be passed almost unnoticed, because St David's station lies away from the city centre and unless one knew just when to look up to the left the twin towers of the beautiful cathedral could be missed altogether. But as the pace quickened again there could be no missing the coastal section from Starcross, beneath the red sandstone cliffs of Dawlish and Teignmouth, with the quaint and picturesque rock formations and the waves occasionally breaking right over the train in rough weather. However much passengers might feel like an after-lunch snooze, no British traveller can resist a sight of the sea,

6. Country station on the Cornish main line: Liskeard, in 1900, with an up train running in. The engine is one of the earlier 5 ft. 8 in. 4–4–0s with domeless boilers, having a straight rather than tapered barrel, and known as the Camel class.

Photomatic

and in such dramatic and colourful surroundings the prospect from the left-hand windows of the train is all-compelling.

Once through Newton Abbot, however, the interest of the ride changes to the more specialised fields of the railway enthusiast, where from the rearward coaches of the train there are frequent glimpses of the head end, swinging round the incessant curves; the exhaust beat of the locomotive is loud, fierce, and rapidly slowing, and the steam from the chimney is shooting almost vertically upwards. On one stretch of the Dainton bank the gradient is 1 in 36, and even with a train of no more than six coaches something near to an all-out performance is needed from a 4–4–0 engine. By that time also they will have been more than $3\frac{3}{4}$ hours on the road, and the fire will not be at its best. In those first years, however, when the 'Limited' ran via Bristol, the engine crews had one advantage over their successors who ran the shortened route to the West, opened in 1906. The lengthy run at slow speed round the Bristol avoiding line gave the fireman a break and an opportunity to get coal forward from the back of the tender and to do some cleaning of the fire. It is true that the engines working the train were furnished with coal of the finest grades, but even so it would not take much in the way of adverse weather conditions to make it a very tough assignment.

The coaches were very pleasant to ride in,

7. The up 'Limited' leaving Penzance in 1906, after the introduction of the massive elliptical roofed Dreadnought coaches. The engine is one of the parallel-boilered Camels, and by that time the track on the viaduct had been changed to the standard cross-sleepered type.

S. C. Nash Collection

8. The down 'Limited' emerging from Parsons tunnel between Dawlish and Teignmouth, and hauled by an Atbara class 4–4–0 no. 3387, 'Roberts'. This picture of 1904 shows the original six-coach set, including elliptical roofed dining-car.

Locomotive & General

though not so ideally suited to sightseeing as the picture-window style of our own times. In the thirds one sat four a side, with not much space between the knees, while in the first- and second-class composites there were a total of six compartments, two for first and four for seconds, each seating three a side. It was perhaps significant of the kind of traffic the Great Western of those days was aiming at that no more than 24 first-class seats were provided on this prestige train, against 128 thirds. It was evidently a step towards what the French call the 'democratisation of speed', and a complete change in policy from the exclusive 'first and second class only' entrée to the old 'Flying Dutchman'. The stock was comfortable without being luxurious but, although relatively light so far as tare weight was concerned, it was later proved rather heavy to pull, from various tests carried out with Churchward's new dynamometer car from 1903 onwards.

The dining-cars run in the new trains were something quite apart from previous Great Western practice, being built out to the extreme width of the loading gauge, no less than $9\frac{1}{2}$ ft wide. They were all of 68 ft long, and had seating for 32 second- or third-class passengers at one end, seated two on each side of the central aisle, and 18 first-class at the opposite end beyond the central kitchen and its equipment. In the dining-car second- and third-class passengers were thus intermingled, while the firsts had the more spacious accommodation to themselves. These dining-cars were the first manifestation of the revolution in coaching stock practice that was taking shape under Churchward's direction; and quite apart from the new look there was the important change in bogie design, which made for a much freer-running vehicle than those with the traditional Dean type bogie.

In Cornwall, as now, no fast running was attempted. For the 53·5 miles between Plymouth North Road and Truro the time allowance was 89 minutes westbound and 90 minutes eastbound. Onwards to Gwinear Road, 15·1 miles, the times were 27 minutes in each direction, and for the 5 miles to St Erth, 8 minutes in the steeply downhill westbound direction and 11 minutes coming up. The final $5\frac{3}{4}$ miles to Penzance were allowed 11 and 10 minutes respectively. The runs between Plymouth and Truro involved some very difficult work, with long ascending gradients up to the bleak, windswept moorlands of east Cornwall, and no opportunity of making fast time downhill because of the unceasing curvature. The maximum speed permitted in Cornwall was officially 60 mph, though this was sometimes briefly exceeded. As will be told later, the Cornish main line had many points of great interest to the railway enthusiast, but to those travellers in the first years of the 'Limited' by the time Plymouth was left behind their only thoughts would have been of journey's end.

9. The up 'Limited' at Coryton Cove, near Dawlish. The section of line through the tunnels between Teignmouth and Dawlish was the last to remain single-tracked. When the 'Limited' was inaugurated the non-stop run between Paddington and Plymouth involved staff exchanging at Parsons tunnel signal box and Dawlish. The engine is an Atbara 4–4–0.
R. S. Carpenter Collection

2

New Route to the West

The 'Great Way Round' to the west, via Bath and Bristol, though a natural link between the original Great Western and the Bristol and Exeter, was not, as sometimes imagined, a monument to Brunel's love of level track and the conservatism of the early management. In the wild days of the Railway Mania plans had been laid for a direct line to the west – the Exeter Great Western – which was to have started at the western end of the Berks section of the Berks and Hants line, at Hungerford, and to have passed through Westbury, Castle Cary, Yeovil, Crewkerne and Honiton to Exeter. But opposition from the London and South Western killed this project, and it was not until the 1880s that revised plans for a shorter route to the west began to take shape. The first stage was in the form of an improved route to Weymouth and involved the construction of a new cut-off line, 13¾ miles long, from a junction with the Berks and Hants Extension line at Patney to the old Wilts, Somerset and Weymouth line at Westbury. The work also involved the doubling and upgrading of the single-tracked Berks and Hants line from Hungerford to Patney, and the building of new intermediate stations to replace the rather antiquated early ones.

The second stage involved the construction of a new cut-off line south-west from Castle Cary to meet the track of the Taunton–Yeovil branch of the Bristol and Exeter, near Langport. However, instead of using the existing connection of this branch to the main line at Durston, which had a very sharp curve, a short additional cut-off was built from Athelney to a main-line junction to be named Cogload, five miles east of Taunton. While the new cut-offs were constructed as first-class fast-running main lines, much of the mileage in the total link-up, particularly on the Berks and Hants section, was anything but suitable for fast express work and much realignment and regrading had to be done, quite apart from the conversion to double track needed between Hungerford and Patney. The contractors for this work, and also for the new Patney–Westbury cut-off line, were Pauling & Company, whose founder was that colossus of railway builders, George Pauling, collaborator with Cecil Rhodes in constructing much of the 'Cape to Cairo' railway in Southern Africa.

Even with the upgrading so assiduously carried out, however, the new route was destined to prove a difficult one in its constant changes of gradient, frequency of curvature and the underlying fact of its being built originally as no more than a very secondary line, without the massive road-bed of the Great Western main line or of the Bristol and Exeter. At Westbury the old station was completely rebuilt, with two island platforms and modern facilities, but in view of the extensive works undertaken it was rather surprising that a better alignment was not obtained for the new main line, coming in from Patney, instead of the sharp curve that entailed a permanent speed restriction of 30 mph. This, however, was rather characteristic of railway practice at the time, and for some years the entry to the Berks and Hants line at Reading West Main signal box was subject to a very severe slack. However, except at Westbury and through the station at Frome, the Wilts, Somerset and Weymouth line was better aligned than the Berks and Hants, though it included some very sharp gradients notably on the Bruton bank. From Westbury the original series of mileposts remained, giving the distances from Paddington via Swindon, Thingley Junction and Trowbridge. These were actually $14\frac{1}{4}$ miles further than those by the new route via

10. The down 'Limited' hauled by a two-cylinder 4–6–0 no. 2905, 'Lady Macbeth' passing beside the Teign estuary just after running through Teignmouth station. The engine is in its original non-superheated condition and the coaches are of the so-called concertina stock.

R. S. Carpenter Collection

Newbury and Hungerford. This new route was opened in 1900 and was used at once in providing an improved service to Weymouth.

It had been hoped to have the Castle Cary–Langport cut-off line ready for the inauguration of the 'Riviera' service in July 1904, but considerable difficulty was experienced in constructing some of the earthworks. To provide an even gradient and facility for fast running in future this new line was heavily engineered through hilly country, and in forming the high embankments and deep cuttings much trouble occurred with slips. It was not until the summer of 1905 that the first part of the new line, from Castle Cary to Charlton Mackrell, was opened, and the remainder was completed early in the following year. The junctions at each end were laid in for express speed, but for some time after the line was opened there were many sections where the earthworks had not fully consolidated and where reduced speed was called for. It ran through a beautiful pastoral countryside, with a distant view of Glastonbury Tor just before reaching Keinton Mandeville station and broad vistas to the north across the Athelney Marshes from Langport onwards. The overall effect of this chain of new lines was to reduce the distance from Paddington to all points west of Cogload Junction by 20·2 miles.

From the inauguration of the summer service of 1906 the 'Limited' was switched on to the new route, and after a short period at 3 hours 3 minutes the standard time for non-stop expresses between Paddington and Exeter became the level three hours, an average speed for 173·7 miles of 57·9 mph. The 'Limited' itself was booked non-stop to Plymouth, first in 4 hours 10 minutes and then in 4 hours 7 minutes but from the outset the three-hour Exeter schedule had its intermediate point-to-point times arranged so as to provide some margin for recovery between Castle Cary and Cogload. The original three-hour timing was as follows:

Miles		Time in minutes	Average speed in mph
0·0	Paddington	0	–
9·1	Southall	11	49·6
18·5	Slough	20	62·6
36·0	Reading	37	61·7
53·1	Newbury	56	54·0
70·1	Savernake	73½	58·3
95·6	Westbury	97½	63·7
108·5	Milepost 122¾	113½	48·4
115·3	Castle Cary	120	62·7
137·9	Cogload Junction	144	56·5
142·9	Taunton	149	60·0
153·8	Whiteball Box	161	54·5
173·7	Exeter (passing time)	180	62·9

These sectional times, if strictly observed, would have required very rapid accelerations from rest and from the various speed restrictions; hard uphill work, up to Savernake, Milepost 122¾, and Whiteball; and very easy running downhill. In later years, when the length between Castle Cary and Cogload became a favourite racing ground, it was easily possible to cut four minutes off the scheduled point-to-point allowance. Initially Rous-Marten expressed the view that even with the new Churchward 4–6–0 locomotives time-keeping would not be possible with more than 300 tons.

The inaugural run of the 'Limited' via Westbury was made on 21 July 1906, with a new 4–6–0 engine, no. 2902, then unnamed but afterwards 'Lady of the Lake'. Rous-Marten recorded that 125 miles were covered in the first two hours, but that then the engine was eased down. The up train was also non-stop from Plymouth to Paddington in 4 hours

11. Newton Abbot station around 1907 or 1908. The Brunel all-over roof can be seen in the background, and the track layout then involved a very severe speed restriction for non-stopping trains. An interesting mixture of old and new coaching stock can be seen.

Photomatic

12. A broadside view of the 'Saint George' – a favourite engine on the Paddington–Plymouth run in the early 1900s. This photograph shows the original style of these famous engines, non-superheated and without top feed. A polished brass worksplate was carried on the centre splasher.

P. J. T. Reed

10 minutes and on the second day, when Rous-Marten travelled by it, the engine was one of the Orleans type French compounds, no. 104.

Coincidentally with the great projects for improving the line there had also been launched the historic development of Great Western locomotive power. Like all Chief Executive Officers of the Company at that time, Churchward attended board meetings, and was on terms of intimate friendship with many of the directors, and he would have been aware of the major aspirations towards the development of holiday business, winter and summer alike, in the West of England. If these aspirations matured to anything like the extent that was hoped, not only would many additional long non-stop runs be required, but there would be much heavier loads to pull. The 'Cornish Riviera Express' as inaugurated in July 1904 could be successfully run by relatively small 4–4–0 locomotives only because its load was very light, limited at first to six coaches. If heavier loads were to be taken at the same standards of speed, it was not merely a matter of providing larger and more powerful locomotives; the coal consumption would need to be kept down to a firing rate that a man could sustain for four hours at a stretch. So Churchward's forward policy became two-pronged: much greater steaming and tractive power; and much lower basic coal consumption. That he

devoted another facet of his overall responsibilities as Chief Mechanical Engineer to an equally important development in coaching stock design was part of the overall policy formulated by the top management.

The master plan for the standardisation of the entire first-line motive power stable of the GWR dated from 1901, and included the large two-cylinder 4–6–0 that eventually multiplied into the Saint or 2900 class for the heaviest express passenger traffic; but experience with the three de Glehn four-cylinder compound Atlantics imported from France in 1903–5 suggested that a still better engine for the long non-stops would be obtained by using four-cylinders with the machinery layout as on the French engines. The celebrated Star class was the outcome – not only an outstanding design in itself, but developed later into the larger Castle and King classes. The prototype, the 'North Star' of 1906, was built as an Atlantic for direct comparison with the French compounds but the first production batch, beginning with No. 4001, 'Dog Star', took up their

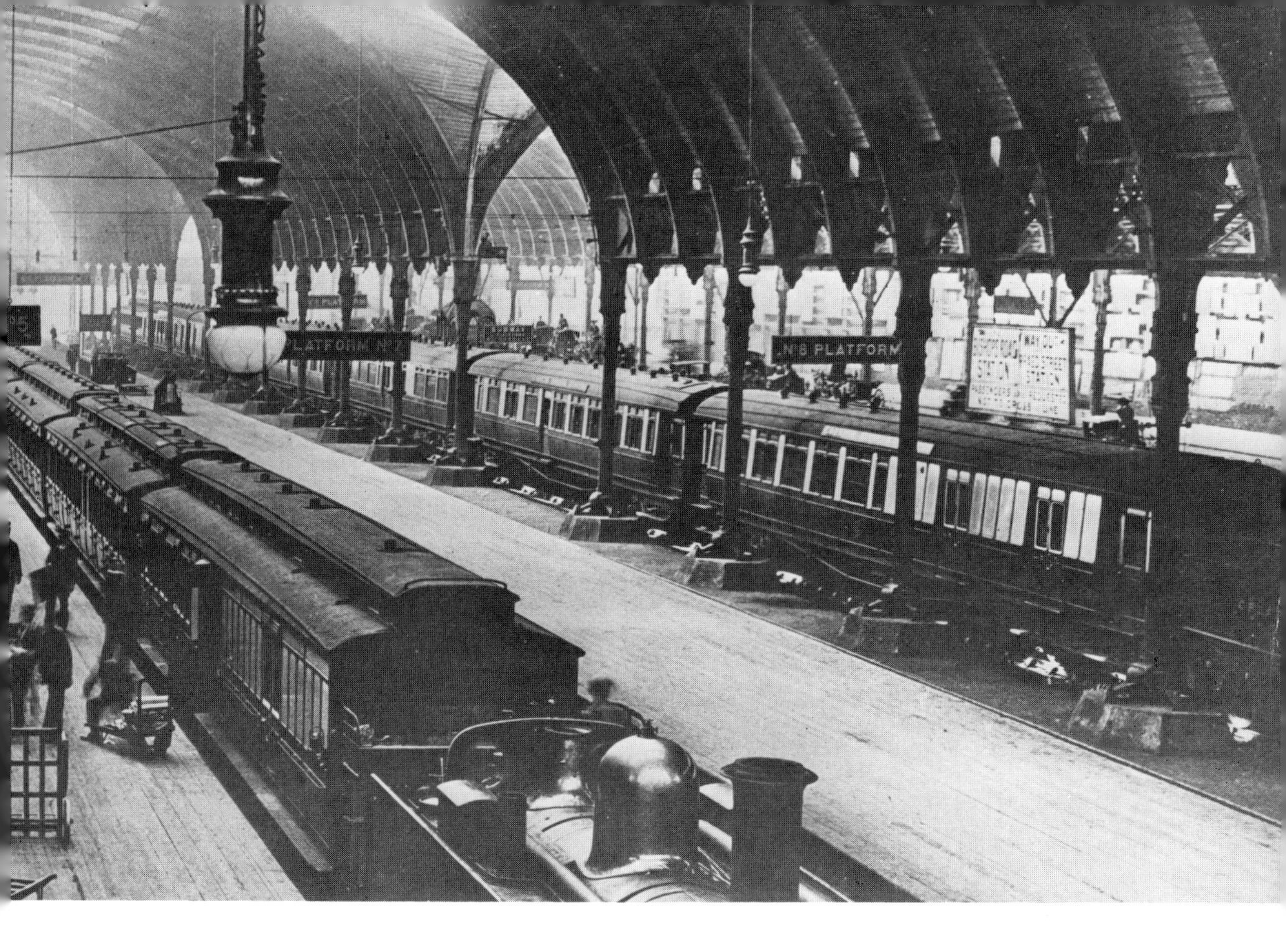

13. Paddington in May 1908. On the right, at no. 8 arrival platform, is the up 'Limited', made up of the Dreadnought stock, at a time when the coach livery was still chocolate and cream. A great variety of other coaching stock can be seen in the station.

Great Western Society

work on the 'Limited' in 1907 and duly earned the warm approbation of C. Rous-Marten, even though he qualified it to the extent of the 300-ton limit of load that he thought inevitable. Nevertheless the two-cylinder engines of the Lady and Saint series also took their turns on the three-hour Exeter non-stops, and also less frequently on the 'Limited' itself. Unlike the original timetabling, which provided for non-stop runs between Paddington and Plymouth in both directions, the eastbound 'Limited' called at Exeter.

I have mentioned earlier that the older Great Western bogie corridor coaches, with the Dean type of bogie, pulled rather heavily in relation to their tare weight, and Churchward's development, using high elliptical roofs and bodies built out to the very limit of the loading gauge, incorporated his own new type of bogie. These coaches were designed to carry the maximum number of passengers for a minimum of tare weight, and to have a low rolling

14. Dawlish station in 1936, looking westwards to one of the most picturesque stretches of the Cornish Riviera route. An up stopping train hauled by a superheated Bulldog class 4–4–0 is leaving for Exeter, with a characteristic mixture of coaching stock.

Ken Nunn Collection

resistance. The earliest type that followed on the design of the restaurant cars used when the 'Limited' was first put on in 1904 were sixty-nine feet long and the thirds seated eighty passengers for a tare weight of thirty-three tons. But although roomy they were not popular. The corridor was on one side of the coach for half its length and then switched over to the other side, and a more serious cause of dislike was that the compartments did not each have individual doors. Out of this first essay Churchward produced his celebrated 70-foot 'concertina' stock. These also were built out to the limit of the loading gauge, but so that the massive door handles did not protrude beyond what was permissible the doors were recessed in, giving the concertina look. These fine cars seated 80 passengers in the third class for a tare weight of only 33 tons.

In comparison with some other British passenger carriages of the period, they were far from luxurious and could, indeed, have been regarded as a little cramped for so prestigious a train as the 'Limited'; but the requirements for a 'Holiday Line', as the GWR was by that time calling itself, were not quite the same as those of top level businessmen on the great lines from London to the north, where a degree of luxury was demanded. Passengers setting out from Paddington for Newquay, Falmouth, St Ives and Penzance would be in holiday mood, and those huge 'concertina' coaches on which the livery had recently been changed from chocolate and cream to crimson lake, lavishly lined out, were certainly impressive. Internally the space beneath the racks was decorated with delightful photochrome coloured views of places on the line, so attractive that many passengers bought copies for home decoration. The roof boards were painted scarlet, although those of the 'Limited' had relatively brief legends compared to some that made their way into the West Country, such as: 'PENZANCE, PLYMOUTH, BRISTOL, SHREWSBURY, CREWE & LIVERPOOL LIME STREET'.

From the time of its transfer to the new route via Westbury, the down 'Limited' carried three separate slip portions: the first for Weymouth, detached at Westbury; the second detached at Taunton, for Ilfracombe; and the third detached at Exeter. In addition to providing fast services to these popular holiday areas it was a logical method of working in that the load on the locomotive was progressively lightened as the gradients became more severe. When Rous-Marten made his first run, with the engine 'Dog Star', the load consisted of five coaches for Penzance and the three slip coaches, representing 300 tons from Paddington, with subsequent reductions of 35 tons each time a slip was detached leaving the engine with a little under 200 tons to take over the South Devon line. It was not long, however, before these loads were being far exceeded. On one journey in 1911, for example, during the summer service when the Taunton slip was not carried, there were eight coaches for Penzance, while the Westbury and Exeter slips each consisted of two coaches, making total loads behind the tender of 415 tons to Westbury, 355 tons to Exeter, and 290 tons beyond. The

15. Planning for a need that did not arise! Churchward's mammoth 'The Great Bear', built in 1908, as a major exercise in boiler design. Although conceived as a development of the Stars, should larger engines be needed on the Cornish Riviera service, this great engine never actually worked on the 'Limited'.
Photomatic

16. The real prototype: the first four-cylinder ten-wheeler, no. 40, built as an Atlantic in 1906 and here seen on the down 'Limited' passing Old Oak Common. This engine, later named 'North Star' was the progenitor of the ever-famous Swindon series of four-cylinder 4–6–0s, Stars, Castles and finally Kings.
Locomotive Publishing Co.

17. The famous 'Cornish Riviera' poster, extensively displayed in years before the first world war.
National Railway Museum

engine was one of the Queen series, no. 4035 'Queen Charlotte', and good time was kept to Exeter. A load of 290 tons was then considered too much for a single engine over the South Devon line and a stop was made at Newton Abbot to attach a bank engine. At that time the working in Cornwall was entirely in the hands of 5 ft 8 in. 4–4–0 engines, used in pairs when the loads were heavy.

If one's interest in the journey was beginning to flag in the early afternoon, it quickly revived on the approach to Plymouth. The approach to tide water at the Laira, the passage of the big engine sheds, where so many Great Western celebrities were usually on display, and the arrival at North Road station were all important, and even the most non-technical of railway enthusiasts would look at his watch on arrival to see whether time had been kept on the world's longest non-stop run. A brief walk up to the front end to watch the changing of locomotives, and then away again on the non-stop run to Truro. For a sightseer a place at the left-hand window or in the corridor is ideal for the first miles out of Plymouth, providing glimpses of the great naval dockyard of Devonport. Then comes that supreme thrill, the crossing of the Tamar at Saltash on the majestic Royal Albert Bridge, passing beneath the entrance arch that is inscribed 'I. K. Brunel, Engineer'. Could there be a more magnificent entrance to the Duchy of Cornwall than this?

Once across the bridge the speed begins to quicken as the train runs beside the tidal creeks of the Lynher River and over the viaduct at St Germans. Here one can see, nearer to the coast, the original alignment of the railway, leading to one of those supremely beautiful timber trestle viaducts designed by Brunel for the Cornwall Railway. But already we are climbing into higher, bleaker regions. The choicest scenery of Cornwall is not inland, on the slopes of the high moorlands, where the derelict and deserted workings of ancient tin mines show up on many a windswept hill crest. At Liskeard one can glimpse from the train the high Moorswater viaduct, and the stone piers alongside that once supported one of the finest of Brunel's timber trestles. Then the train begins to descend through Largin Woods to one of the most beautiful stretches of the Cornish main line, coming eventually to Lostwithiel.

I shall always remember my own first journey on the train. It was in late summer and the skies were clouding over before we left Paddington. By Taunton it was raining, and we ran through South Devon, Plymouth, and across the Tamar in a thick unseasonable drizzle. By Lostwithiel the rain had certainly ceased, but clouds were so low as to be touching the heights above the little town, and our engine was slipping slightly on the stiff climb to Treverrin tunnel. But then there began a marvellous transformation. Across the sands at Par Harbour we could see the skies beginning to clear, and as we pounded up the steep gradient towards St Austell there were already patches of vivid blue. Past the conical dumps of the china clay workings, on over the switchback grades, and when we rolled into Truro sunshine was sweeping the platforms. The colouring even in the scarred tin-mining regions became brilliant, and after the brief stop at Gwinear Road we rode steeply downhill for an enchanting distant sight of St Ives set against a brilliant sea and sky; and so finally past Marazion, with that vista of St Michael's Mount that one knew so well from childhood from countless pictures, and along the foreshore and into Penzance.

The Cornish main line is not a place for the seeker after high speed and spectacular point-to-point averages. It is a case of hard slogging

18. The up 'Limited' entering Paddington in 1910, hauled by one of the King series of four-cylinder 4–6–0s, no. 4028, 'King John'. This engine was then in its original condition, non-superheated as built in 1909 and without the familiar top feed apparatus on the safety valve bonnet.

Ken Nunn Collection

up the banks, usually falling to well below 30 mph at the summits, followed by steady descents, rarely touching as much as 60 mph. The actual gradients are not so steep as the worst pitches of the South Devon line, and locomotives can take heavier loads without a pilot. In the early days of the service, before the allocation of many of the Mogul engines of the 4300 class to Cornwall, one often saw one of the domed-boiler Duke class 4–4–0s piloting a Bulldog, though after the First World War the Moguls had the Cornish main line almost to the exclusion of all other types. As the popularity of the Cornish holiday resorts grew the load of the 'Limited' was increased by inclusion of through carriages for Falmouth and St Ives, detached at Truro and St Erth respectively, and at times an extra coach was included for Plymouth passengers. Eight 70-foot coaches became the maximum taken unassisted by the Star class engines over the South Devon line. The official tonnage limit was 292 because it had been found that since superheating those splendid engines could take a heavier load than in the early days of the service, when only the non-superheated engines of the original Star and Knight series were available. Rous-Marten's forecast of 300 tons throughout was to be still more dramatically given the lie in the years 1912–16.

3
Wartime and Recovery

The outbreak of war in August 1914 at first affected the West of England lines of the Great Western less, perhaps, than most other areas of the British railway network. For, while Portland and Devonport were great centres of naval engineering and ship maintenance, the Grand Fleet itself and all its vast personnel was based far away in the north of Scotland. Furthermore, there was a popular sentiment that, war or no war, little should be allowed to interfere with holidays in Britain. While the east and south-east coasts became subject to threats of invasion and of coastal raiders, Devon and Cornwall were thought to be relatively safe. In consequence passenger traffic became heavier than ever. In the earlier stages of the war no one thought of decelerating the trains, and the tasks set to the locomotives were increasingly heavy. Even before the summer of 1914 there had been occasions, fully documented in *The Railway Magazine* and elsewhere, of engines of the Saint and Star classes conveying loads of around 500 tons, without pilot assistance, so the locomotive department was prepared for trains that would have barely been thought of when the 'Limited' was first put in.

By that time enough of the four-cylinder 4–6–0 locomotives had been built for them to be used to the exclusion of all others in the Paddington–Plymouth top links. The addition of superheating increased their already high capacity and on the last twenty of the class, named after princes and princesses, the diameter of the cylinders was increased from 14¼ to 15 in. The working of express trains in the height of the holiday season frequently led to the 'Limited' and other popular services being run in two or more sections, and then the slip portions were run in the second or third division. In the case of the 'Limited' itself the first section would be for Cornish stations only, while the second would be run non-stop to Exeter and would convey the Westbury and Taunton slips. In these conditions the first part would not get the successive reductions of load that were so helpful to the enginemen on the non-stop run to Plymouth and the full load for Cornwall would nearly always require the attachment of a bank engine from Newton Abbot westwards.

A remarkable instance of working in similar circumstances was fortunately documented fully by an experienced recorder of train running. The load from the start was thirteen coaches, of which eleven had to be conveyed to Truro. From Paddington the engine was of the 15-inch variety, no. 4045 'Prince John',

19. Advertising for the Cornish Riviera in a London square, just after the first world war. The road vehicle, with its solid rubber tyres, was also typical of the equipment of some Great Western bus services in Cornwall, to beauty spots beyond the reach of the railway.

British Rail

with 470 tons behind the tender. Despite very good work exact point-to-point times could not be kept at first; Westbury was passed in $98\frac{1}{4}$ minutes, $\frac{3}{4}$ minute late, and still with the full load another $1\frac{1}{2}$ minutes were lost on the difficult section to Castle Cary. But by the summer of 1914, when the run was made, the Langport cut-off line had fully consolidated and the driver took full advantage of the favourable gradients and good alignment to run really hard. Instead of the scheduled 24 minutes he took only 19 minutes for the 22·6 miles, and so was able to pass Taunton $3\frac{1}{4}$ minutes early, and continue still with the full 470-ton load. To climb up to Whiteball tunnel in these conditions required a virtually all-out effort. The loss on the scheduled point-to-point time was fractional only, and with fast downhill running afterwards the two-coach slip portion was detached at Exeter while the main train ran through in $177\frac{3}{4}$ minutes from Paddington. This was a magnificent piece of enginemanship on the part of driver and fireman.

At Newton Abbot a stop was made for assistance, and a 4–4–0 of the Duke class coupled on ahead of the 'Prince John'. The run of 32 miles on to Plymouth was made in a few seconds under 43 minutes and the train arrived in $245\frac{1}{2}$ minutes from Paddington, $1\frac{1}{2}$ minutes early, inclusive of standing $2\frac{1}{4}$ minutes at Newton Abbot. The eleven-coach train

20. One of the most elegant Great Western coach styles of all time: the 'concertina' stock, pictured after the restoration of the chocolate and cream livery in 1922.

British Rail

which had to be conveyed forward to Truro weighed 395 tons and was taken by one of Churchward's new 5 ft 8 in. Mogul engines of the 4300 class, some of which had already been allocated to the West Country. This engine did well to cover the 53½ miles from Plymouth to Truro non-stop in 77¾ minutes, 3¼ minutes inside schedule time. It was, of course, not to be expected that work of such quality would be forthcoming during the war years, either east or west of Plymouth; but a tradition of service had been established that the top link drivers at Old Oak Common, Newton Abbot and Laira sheds made it almost a point of honour to maintain. It was perhaps inevitable that the Laira men became the supreme specialists in working the 'Limited'. At Old Oak Common the top link men ran to Wolverhampton, to Cardiff, and to Bristol as well as to the West Country and were more in the way of general practicians, whereas the Laira men had the difficult South Devon line at their doorstep.

In the last year of the war in which the 'Limited' ran to its pre-war schedule, 1916, two magnificent runs were recorded with engine no. 4018, 'Knight of the Grand Cross', starting out of Paddington with fourteen coaches on the first occasion and fifteen on the second – shades of Rous-Marten's ideas of a 300-ton maximum! I have set out below the actual times as far as Exeter. On the first run the recorder was travelling only to Torquay, and the time at Exeter is that of arrival of the slip portion. On the second the record was continued to Plymouth. Because of checks both trains passed Savernake behind time,

THE 'LIMITED' IN 1916: TWO RUNS WITH ENGINE 4018

Load to Westbury			14 – 490 tons	15 – 535 tons
,, ,, Taunton			11 – 400 ,,	12 – 435 ,,
,, ,, Exeter			9 – 320 ,,	9 – 340 ,,

Distance in miles		Scheduled time in minutes	Actual time in minutes	Actual time in minutes
0·0	Paddington	0	0	0
–		–	signal check	–
36·0	Reading	37	42¼	41
–		–	–	permanent way check
70·1	Savernake	73½	80¼	81
95·6	Westbury	97½	102¾	104
115·3	Castle Cary	120	125¾	127
137·9	Cogload Junction	144	144½	146½
142·9	Taunton	149	149¼	151
153·8	Whiteball	161½	161¾	162½
173·7	Exeter	180	179½	180

21. One of the most famous of Cornish Riviera engines, 'Pendennis Castle' here seen leaving King's Cross (LNER) during the historic interchange trials of 1925, when the engine worked with great distinction between King's Cross and Doncaster. It is now in North-Western Australia, on the mineral railway of the Hamersley Iron Company (Pty) Limited.

Rail Archive Stephenson

and little if anything could be regained with these heavy loads until Castle Cary had been passed. Then one can note the very fast running over the Langport cut-off line, and an actual gain of a minute on the second run between Taunton and Whiteball summit. Both trains passed Exeter on time and on the second, continuing to Plymouth with 265 tons, the arrival there was a minute early.

In wartime such running could not be sustained and in January 1917 the 'Limited' was suspended. A substitute left Paddington at 10.15 a.m., carrying no slip coaches or restaurant car and stopping at Westbury, Taunton and Exeter. The arrival time at Penzance became 6 p.m. instead of 5 p.m. – an increase in journey time of $1\frac{1}{4}$ hours. The fastest time between Paddington and Exeter became $3\frac{1}{2}$ hours. All this was necessary, together with the withdrawal of many supplementary services, to save coal and minimise the need for maintenance work on track and rolling stock. The once gay livery of the express locomotives disappeared and they went about their work in plain green, devoid of any lining out, with all the bright parts painted over. Generally speaking the standards of maintenance were not impaired and occasionally when there were exceptional demands fine work was done, though from January 1917 the maximum speed over the whole line was 60 mph. The 'Cornish Riviera Express' did not remain suspended

22. What the fireman saw: from the footplate when climbing the very severe Dainton incline, South Devon. The signal box ahead is Stoneycombe, controlling the connections to the neighbouring quarries.

Kenneth Leech

23. One of the smooth-sided 60-ft corridor-thirds, used on the 'Limited' in the late 1920s. The elaborate lining out of earlier days had been much simplified by then. Note also the panels at each end stating the seat numbers to be found in this particular coach. At busy times nearly every seat on the 'Limited' would be reserved in advance. British Rail

for very long, however. It was restored less than a year after the Armistice of 1918, for the summer service of 1919, leaving Paddington once again at 10.30 a.m. and again running non-stop to Plymouth. The 60 mph limit remained and because of this full pre-war speed could not be restored between Paddington and Exeter. An extra 15 minutes were put into the schedule on this account, making the passing time there 1.45 p.m. instead of 1.30; but from Exeter westwards the times were exactly those of pre-war days, with Plymouth reached at 2.52 p.m. instead of 2.37 and Penzance at 5.15 p.m. instead of 5.00.

The 60 mph limit was a great handicap when it came to making up time lost by signal and permanent way repair checks, and a run made by one of the other West of England expresses on the $3\frac{1}{4}$-hour non-stop schedule from Paddington to Exeter shows that time could barely be kept, even with a slight 'bending of the rules'. The train was the 11.30 a.m. from Paddington with a load of 395 tons, hauled by a famous Plymouth engine of the day, no. 4038 'Queen Berengaria'. Adverse signals and a long check for permanent way repairs in the early stages caused the first 70·1 miles, to Savernake, to take $82\frac{1}{4}$ minutes, a loss of 5 minutes. Then the two usually very favourable lengths of 25·5 miles down to Westbury and 27·6 miles between Castle Cary and Taunton took $24\frac{1}{2}$ and $27\frac{3}{4}$ minutes respectively – demonstrating fairly close adherence to the 60 mph limit. It was only when the train was still behind time on passing Whiteball summit that the driver was tempted to make a brief spurt at 70 mph to bring the train into Exeter only a minute late.

The restoration to full pre-war standards of speed came in the autumn of 1921, and this was signalised by some magnificent running on occasions. There was one classic, when fortunately the late Cecil J. Allen was a passenger, when four permanent way checks caused the train to pass Castle Cary 10 minutes late on the accelerated schedule. So vigorous was the subsequent running that Exeter was passed less than a minute late and Newton Abbot $\frac{3}{4}$ minute early in $202\frac{1}{4}$ minutes for the 193·9 miles from Paddington. In striking contrast to the run of 1920 with 'Queen Berengaria', the 27·6 miles from Castle Cary to Taunton took only $22\frac{3}{4}$ minutes with much running at 75 or 80 mph. The engine was no. 4003, 'Lode Star', the member of the class that is now preserved in the Railway Museum at Swindon.

The restoration of full pre-war speed on the 'Limited' together with all the various through carriages, slip coaches and restaurant cars was accompanied by a resurgence of publicity for holidays in the Cornish Riviera. It so happened that the autumn of 1921 was marked by a spell of very fine weather. I remember it well because it was my first term at Imperial College in London, and the hot sunny days that lasted well into November enabled me to do some railway photography on the lines around London, while at Paddington new posters appeared showing bathing parties in Cornwall, where an Indian summer was then prevailing, in November, together with an exhortation to indulge in late holidays. It was at that time also that the Great Western entered into partnership with the Great Central, the North Eastern and the North British for running a through carriage service between Penzance and Aberdeen from 3 October 1921, and, although in neither direction were the through carriages attached to the 'Limited' itself, full opportunity was taken at an informal dinner at the Queens Hotel, Penzance, by representatives of the GWR to extol the holiday attractions of the region.

The restoration of full pre-war speed and facilities on the 'Limited' in 1921 was accomplished with locomotives and coaching stock of pre-war build. The engines were still in plain, unlined green – though very smartly turned out – while the carriages were finished in crimson lake, with scarlet roof boards. Early in 1923, however, the Great Western introduced a new type of bogie corridor 70-foot coach. It was the first manifestation of the change in Chief Mechanical Engineers, when C. B. Collett had succeeded Churchward in that distinguished office as from January 1922. Externally the new carriages, which were put at first on to the South Wales service, could be readily distinguished by their smooth sides, quite devoid of the panelled effect on the older varieties, though they were lavishly lined out. The feature that attracted by far the most attention from railway enthusiasts, however, was the restoration of the historic coach livery of the GWR, dating from broad gauge days, namely the ever-famous chocolate and cream. Because of the greater use of steel in their construction and increased strength against buffing shocks, they were considerably heavier, with the ten-compartment thirds turning the scale at 38 tons.

The introduction of these coaches in the chocolate and cream livery was followed at once by the repainting of all other passenger vehicles similarly as they passed through Swindon works for overhaul; but the locomotives

24. Rival of the Castles in 1925: the Gresley Pacific engine no. 4474, later named 'Victor Wild', leaving Paddington. This engine did some notable work on the 'Cornish Riviera Express', keeping good time with maximum load trains.
Rail Archive Stephenson

Nº 447

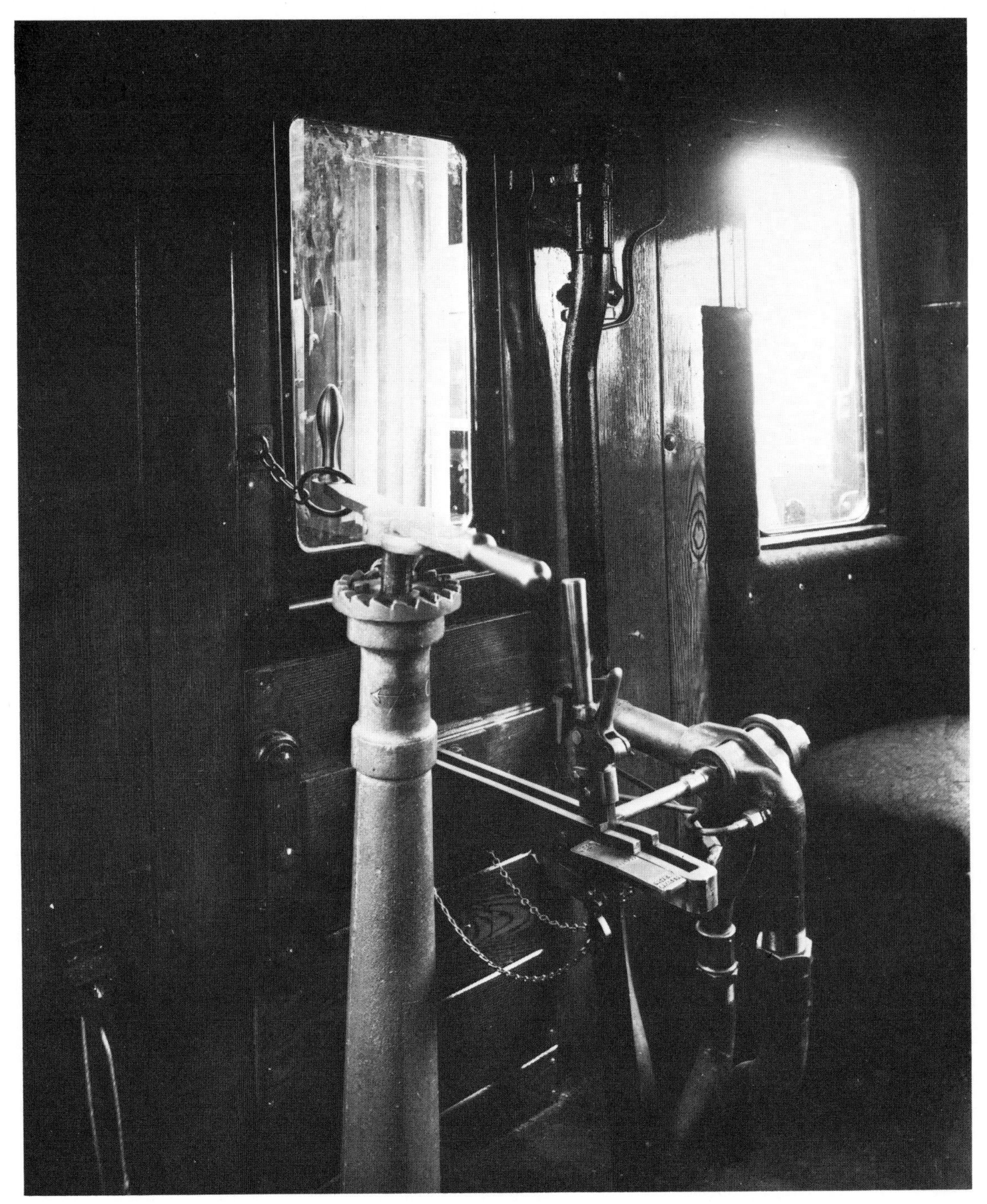

25. The 'slip coach' was a feature of Cornish Riviera operation: no fewer than three portions were detached, at Westbury, Taunton and Exeter. This picture shows the very simple apparatus provided for the guard in each of the slip portions. British Rail

themselves remained in plain green during the summer of 1923. In 1922 a further batch of Star class engines named after abbeys had been put into service, and these like many other engines of the Saint and Star classes at that time had plain tapered cast-iron chimneys instead of the large built-up type with copper tops, so characteristic of the Great Western in pre-war years. Those chimneys added to the austere look of the locomotives in the war years and just after. When the new steel-panelled coaches were put on to the 'Limited' the load became heavier than ever, but the Stars took fourteen out of Paddington without any complaints from their crews, and the gross loads behind their tenders were frequently in the region of 520 or 530 tons. This of course took place only at peak periods, which meant the early summer, while the full complement of slip coaches was being carried, and in the period just after the end of the summer service. During the height of the summer the 'Limited' carried only the Westbury slip portion; services to Ilfracombe, Minehead and the Torquay line were provided by entirely separate trains.

The year 1923 was distinguished still further by the introduction of the Castle class of locomotive. The GWR management was nothing if not publicity minded, and the fact that the new engines had the highest nominal tractive effort of any British passenger locomotive of the day led to a tremendous campaign of one-upmanship, particularly as the nearest rivals of the Castles were the very much larger and heavier Pacifics of Gresley and Raven design on the newly formed London and North Eastern Railway. Until the end of 1922 the Great Western had taken pride in having the only Pacific engine running on British railway metals, and amongst the northern rivals of 'The Great Bear' there was much comparing of dimensions and outward characteristics to make sure that Great Western supremacy had actually disappeared. Even the tractive efforts of the new engines were greater. But when 'Caerphilly Castle' took the road, with 31,625 lb. of tractive effort against the 29,835 lb. of

26. The 'slip' slipped: a 1956 photograph taken from the Westbury slip coach just after it had been detached from the main train. After the construction of the Westbury by-pass line the slip was detached, often at well over 70 mph, at Heywood Road Junction, as shown here, and subsequently hauled into the station by a local pilot engine. H. C. Casserley

the Gresley Pacific, it was clear that the GWR was definitely in the lead again, even though the new engine was a 4–6–0 and not a 4–6–2. Of course they were put on to the 'Cornish Riviera Express' – by some way the hardest express duty on the line, and probably the hardest in all England at that time.

The names were another fine stroke of publicity, but in choosing castles I wonder if anyone at Paddington or Swindon recalled that the suggestion for such a class had been made in *The Railway Magazine* as far back as 1908? In November of that year C. Rous-Marten voiced his disapproval of the choice of flowers for a new series of double-framed 4–4–0s:

> Many sources of nomenclature of a more appropriate character might quite easily have been drawn upon. Besides the names of old broad gauge locomotives, which include most of the more famous 'fathers of the locomotive' there are available the names of novelists, men of scientific attainments, modern statesmen, and the like; or why not a 'Castle' class? There are, I believe, only four Great Western Railway locomotives now running which bear the names of castles, and there are surely more castles than this within the railway's system?

Four there certainly were in 1908: 'Wigmore Castle' and 'Windsor Castle' on 4–2–2 singles, scrapped in 1909 and 1913 respectively, and 'Pendennis Castle' and 'Chepstow Castle' on Duke class 4–4–0s. Both the last named were taken for new 4–6–0s in 1923 but so far as the 'Cornish Riviera Express' was concerned, of the ten names chosen for the first batch of Castles only two, 'Powderham' and 'Pendennis', had a West Country flavour. Of the rest, six were Welsh, or Welsh border country, and two English.

Again, however, to railway lovers it was the glittering turnout of the Castles more than their names and technical details that so appealed to the senses, because all the glorious colouring of old, with a wealth of shining brass and copper work, was restored. More than this, engines of the Saint and Star classes were similarly bedecked once again, as they went through Swindon for overhaul. All of the first batch of Castles were stationed at Old Oak Common, which meant that they were all available to haul the 'Limited'. It did not matter much if one had an engine with a Welsh name; the sight of a highly burnished Castle at the head of a long train of 70-foot chocolate and cream coaches was the most satisfying spectacle a Great Western enthusiast could desire. On alternate days Plymouth engines and men would be on the job, and it seemed as though the Laira crews with the smaller engines definitely set out to equal, or even try to surpass, the achievements of the London men with the Castles. The work of 'Royal Star', 'Knight of the Thistle', 'Queen Adelaide', 'Malvern Abbey' and 'Reading Abbey' in the years 1923–5 was superlative by any standards.

All this, however, was building up towards one of the classic locomotive contests of all time, fought out on the 'Cornish Riviera Express'. In the spring of 1924 one of the best known runners of Old Oak shed, no. 4074

27. A corner of the first-class dining-car, in the days when the full lunch on the 'Limited' cost three shillings and sixpence! British Rail

GWR
GWR

'Caldicot Castle', and her driver, F. Rowe, ran a series of trials with the dynamometer car. These were conducted from Swindon to Plymouth and back, with trains made up to the heaviest tonnages permitted over each stage of that difficult route. Subsequently C. B. Collett published the results in a fascinating paper read before the World Power Conference of 1924. The results, expressed in terms of the power exerted and, more important still, of the coal consumed in relation to the work done, were so good that his fellow chief mechanical engineers just did not believe them. Urgent studies took place on the LMS for example, comparing the proportions of the Castle with that of the Lancashire and Yorkshire type 4–6–0, for which they had recent authoritative test results; and the efficiencies were so different, and so heavily in favour of the Castle, that the men of the LMS seriously doubted their veracity. The disbelief of the LNER locomotive department was to take a more active form in a year's time.

4

Locomotive Pre-eminence

Amid the intense publicity that was showered upon the 'Cornish Riviera Express' following the introduction of the Castle class locomotives, it must be emphasised that the train itself had three distinct focal points of interest and affection, two of which were quite unconnected with the form of motive power. The first was that so successfully fostered and nourished by the publicity department of the GWR – that the train was the essential beginning of a grand holiday in the West of England. On the platform at Paddington, the splendid 'open' No. 1 from which the 'Limited' always used to depart, one could buy for a mere shilling booklets describing the scene 'Through the Window'; the dining-car conductor was out on the platform booking seats for lunch – and one needed to book for there were often three sittings! – and in those far off days the full table d'hôte meal cost three shillings and sixpence. Those who were interested in a little more than their own destination would see through-carriages for many other holiday resorts in the West Country. The holiday feeling was everywhere along no. 1 platform at Paddington, with the leading coaches out of sight round the curve at the head end.

The second point of affection was that of the Cornish people, particularly those in the far west. In Penzance it was always referred to as the 'Riviera'. When the holiday season was nearly at its end, tradesmen, hoteliers and others partook of the holiday spirit in reverse, and travelled on the 'Riviera' to London. No one in Penzance thought of going by any other train, or indeed any other means of conveyance, in those days. Artists working in Newlyn and St Ives sent their pictures to the London exhibitions by the 'Riviera' and, although there were special trains in the appropriate seasons for spring flowers and choice vegetables, small private consignments to friends east of the Tamar were loaded personally into the train's luggage vans. The one thing noticeably absent from the clientele of the train was the hustling businessman, who patronised the Great Western expresses from London to Bristol, South Wales and the Midlands. The men to whom time was precious could not afford to use the day-long run of the 'Limited'.

Then of course there were the railway enthusiasts, and particularly the locomotive enthusiasts. With the gradual increase in its load to the normal winter formation of fourteen coaches, maintaining the same overall running times as those established in 1907

28. Holiday rush at Paddington: engines of three sections of the 'Limited' abreast at platforms 1, 2 and 3, and carrying the identification numbers used at times when many extra trains were running. The engines, left to right, are an unidentified Castle and the Kings 6012, 'King Edward VI', and 6001, 'King Edward VII'.
British Rail

when the Exeter run was shortened to the even 3 hours from the 3 hours 3 minutes of the first service via Westbury in 1906, the westbound train became the hardest daily task set to any British locomotive. The tonnages behind the tender were 530 to Westbury, 450 to Taunton, about 380 to Exeter and 310 to Plymouth. At that time the train went through Westbury and Frome stations, with a 30 mph speed restriction at each place, and at first there was always a very severe slowing through Newton Abbot, where the station was being rebuilt. It must be confessed that not even the new Castle engines always kept strict time between Paddington and Westbury with the full load. Sometimes two or three minutes would be dropped on the hard schedule of $97\frac{1}{2}$ minutes for that initial 95·6 miles, to be regained by fast running over the Langport cut-off line. The curious thing was that there was no corresponding difficulty on the eastbound journey. One could not perform the slip coach act in reverse, and the up 'Limited' had its full load on leaving Plymouth. It was a much easier train from the locomotive point of view.

The difficulty of working the down 'Limited' with its full winter load became highlighted

29. Five minutes before departure time: no. 1 platform at Paddington, 10.25 a.m., with an animated scene alongside the 'Limited'.

British Rail

at the time of the famous locomotive interchange trials in 1925. At the British Empire Exhibition at Wembley in 1924, engines of the Castle and Gresley Pacific types formed the respective centrepieces of the Great Western and London and North Eastern stands, in close proximity to each other; and for those who read the not so very small print describing these exhibits the Castle was quoted as the most powerful passenger locomotive in Great Britain. With the much larger Gresley Pacific nearby, some found this rather hard to believe. How the interchange trials of the following year were arranged has never been fully established, but one thing I can make clear: there was never any question of a challenge from the locomotive department of the London and North Eastern to the Great Western to prove the superiority of the class. In fact Mr Gresley, as he then was, was rather reluctant to embark upon it, because his Pacific design was not then fully developed. The trials were initiated at higher management level than the respective locomotive engineers. That the Gresley Pacific kept the schedule of the down 'Limited' on her very first trip was an astound-

30. A famous Cornish viaduct: Moorswater, near Liskeard, showing the branch line to Looe passing underneath and the masonry piers of Brunel's original viaduct, carrying a single line, beside the massive yet graceful arches of the present structure.
J. Spencer Gilks

ing piece of enginemanship, having regard to the exceptional difficulty of the job for the regular experienced crews of Old Oak Common and Laira sheds.

Unhappily, the Great Western management chose to publicise the results of what was intended to be a wholly friendly and co-operative affair in a blatant spirit of one-upmanship, which disquieted their own supporters as much as it infuriated the LNER. It is true that the Castles seemed to have had the best of it during the week's running, both on the 'Limited' and on the exchange workings between King's Cross and Doncaster. But to publish the results unilaterally was tactless, to say the least of it, and it put the Great Western and its locomotive department somewhat outside the comity of the British railway engineering world for some years to come. National events also tended to show that the superiority of the Castle engines suggested by the published results of the trial week in 1925 was not sustained in all circumstances. In 1926 the General Strike and the prolonged coal strike that continued throughout the summer and into the autumn of that year

31. The continuing Swindon tradition: the enlargement from the Star of 1907, to the Castle of 1923 and the King of 1927. The engines are no. 4004, 'Morning Star', as superheated, no. 5010, 'Restormel Castle', and no. 6000, 'King George V'.

Locomotive Publishing Co.

32. Advertising the 'Limited' in the USA: reproduction of a splendid coloured poster, displayed in America after the Coronation of 1937, when engine no. 6028, originally 'King Henry II', had been renamed 'King George VI'.

British Rail

resulted in the importing of much foreign coal, and bereft of their usual supplies of high-grade Welsh coal the Great Western top link express locomotives seemed very much shorn of their Samsonian locks. Much of the express train running in 1926, even with the keenest and most experienced engine crews, was far below normal standards and involved occasional losses of time.

The scrapping of the first British Pacific engine, 'The Great Bear', in 1924 had repercussions that came to affect the Cornish Riviera route to a marked extent. The great engine had never at any time worked on the train, but by dismantling it and replacing it with a Castle carrying the same number, Collett dived into a hornet's nest. At Paddington, the General Manager, Sir Felix Pole, wanted to know why this prestige symbol had been scrapped and why it was so restricted as to its route availability. The civil engineer had to explain how there were certain bridges on the West of England route that would not take its weight. Then, to add fuel to the fire, in 1926 the Southern Railway built the first of the Lord Nelson class 4–6–0s which, having a nominal tractive effort greater than the Castle, was promptly claimed as Britain's most powerful locomotive. By that reckoning it certainly was! This was too much for Sir Felix Pole. He told the civil engineer to hurry forward the strengthening of those bridges, and told Collett to design and get built in time for the summer traffic of 1927 a 'super' new locomotive that in nominal tractive effort would far outstrip the 'Lord Nelson'.

The new locomotive design, planned as the largest possible enlargement of the basic Churchward Star of 1907, was naturally intended for the West of England route, but other circumstances ordained that the original target was not Plymouth North Road but Baltimore, in the USA. Sir Felix Pole had promised that a Great Western engine should be sent over for the centenary celebrations of the Baltimore and Ohio Railroad in August 1927, and the ever-famous 'King George V' had time for no more than a single run down to Plymouth with the 'Limited' before it was partially dismantled and then shipped across to America. But the great engine had hardly arrived there, to be the wonder and delight of all who saw it, when near-disaster struck the 'Limited' at home.

In the twentieth century the Great Western had built up a wonderful safety record. This was in large measure due to the care taken in design, the vigilance in maintenance, and above all the high sense of responsibility shown by all members of the staff concerned with train running. There was also undoubtedly an element of luck, in that when mishaps did occur the circumstances were on the side of safety rather than disaster. This was certainly the case at Midgham on the morning of 10 August 1927.

'King George V' had first worked the train on 20 July, and since then the remaining five engines of the first batch of Kings had gone into regular service on the Paddington–Plymouth route. Although acknowledged as very powerful and free-running engines, there were soon reports of unpleasant rolling at high speed. This was something new for the Great Western. A Star or a Castle might give an isolated lurch on a weak rail joint; but the Kings seemed prone to something more sustained and disconcerting. On the morning of 10 August, when engine no. 6003, 'King George IV', was running at about 60 mph with the down 'Limited', the bogie became derailed. It was fortunate beyond measure that this mishap occurred on plain line, and that the driver was able to stop without further incident. Had there

33. The up 'Limited', with the 'Centenary' coaching stock of 1935, passing Coryton Cove, Dawlish, hauled by engine no. 6007, 'King William III'. This picture may be compared with that of the 'Limited' in 1904 at the same spot, hauled by an Atbara class locomotive, on page 21. R. S. Carpenter Collection

been points or crossings in their path nothing could have prevented a bad accident. Naturally a very serious view was taken of it, and the most urgent action was taken to ascertain the root cause and to redesign the springing of the bogie to eliminate any such risk in future. These measures were wholly successful, and thereafter – as I can verify from several thousands of miles travelled on their footplates – the Kings became among the steadiest and best riding of all British express locomotives. But Midgham was a mighty close shave!

The introduction of the King class locomotives with their greatly increased tractive power was followed at once by cutting the journey time to Plymouth to the even 4 hours, instead of the time-honoured 4 hours 7 minutes that had prevailed since 1907. An even more significant improvement was that the limit of load for an unpiloted engine between Newton Abbot and Plymouth was increased to 360 tons. During the time of heavy summer traffic this eliminated many of the stops for assistance over the South Devon line that had previously been necessary with the smaller engines. The increasing popularity of Newquay as a holiday centre led to a change in the working in Cornwall. While there were through services in the summer holiday season, the 'Limited' during the winter months served Newquay by a connection from Truro. From the autumn of 1927, however, advantage was taken of the earlier arrival at Plymouth to insert a stop at Par and to detach a through carriage for Newquay at that junction. In the early 1920s when I first travelled on the 'Limited' the motive power in Cornwall was provided by the standard Moguls, not quite to the exclusion of all others but very nearly so. The up 'Limited' was sometimes worked by a Saint class engine from Exeter shed between Truro and Plymouth, and the West of England postal special

34. Even the Kings were not permitted to haul a heavier load than 360 tons over the fearsome gradients of the South Devon line. A stop had to be made at Newton Abbot for a bank engine, as in this case where a Bulldog, the 'Jupiter', is assisting the train engine up Dainton bank.

John Ashman FRPS

35. Frontispiece from the 1939 edition of *Through the Window*, a paperback describing points of interest on the route from Paddington to Penzance.

National Railway Museum

nearly always had a 4–6–0. But by the beginning of the 1930s the Hall class were in full possession. Nevertheless the introduction of the Kings released a few Castles for service in Cornwall, and one of these was usually to be seen on the 'Limited' in each direction.

The load limit for those engines was much higher in Cornwall than over the South Devon line, and I have a record of no. 5028, 'Llantilio Castle', having to take a gross load of 410 tons westwards from Plymouth, without assistance. The Castles were definitely not at their best in slogging up 1 in 60 gradients at between 25 and 30 mph, and on the Cornish banks the Halls, with Churchward's setting of the Stephenson link motion, did almost as well, despite their lower nominal tractive effort. There was no chance of fast and free running in the true Castle style in Cornwall, and when occasionally a venturesome driver did bend the rules a little and gave us brief downhill spurts at 65 or 70 mph the riding in the train could be distinctly lively! I remember vividly what heavy weather 'Llantilio Castle' made of the gradient from Par Harbour up to St Austell, falling to 21 mph with much noise from the exhaust, while we in the train were out in the corridors enjoying the sight of a wild Cornish sea racing in to smash against the cliffs of Carlyon Bay.

Nevertheless, whatever passengers and railway enthusiasts might think of it, the early 1930s were anxious times for the GWR. The great slump probably affected the company less than it did the other British main-line companies and the funds made available by government loans, at advantageous rates of interest, were spent partly on improvements to other parts of the line. Important works that helped in the running of the 'Limited' were the construction of the by-pass lines avoiding the junctions and curves at Westbury and Frome. These eliminated the hampering speed restrictions at those points, and enabled full speed running to be maintained through from the straight western end of the Patney and Westbury cut-off line of 1900 to the Wilts and Somerset line south-west of Frome. This work not only enabled faster time to be made, but effected a reduction in coal consumption on the journey by avoiding the vigorous accelerations that had previously been necessary to regain speed after those two 30 mph slacks. Slump or no slump, however, in 1935 the Great Western had to celebrate the centenary of its Act of Incorporation, and this was signalled by, among other events, the introduction of new rolling stock on the 'Cornish Riviera Express' – very wide and, significantly in view of future trends, with entrances only at each end.

At this time also the Board of the GWR were becoming increasingly worried about the price of locomotive coal, and despite the high economy in working of Kings, Castles and indeed of all other Great Western main-line locomotives, they began seriously to consider electrifying at least part of the main-line network. The principal source of supply was, of course, South Wales, and it was not only the cost of coal at the pit-head that was in question but the job of carrying it to the sheds where it was needed. With those two considerations in mind an approach was made to a famous firm of consulting engineers to work out a scheme for electrifying the whole line, branches as well as main tracks, west of Taunton. Electric traction is particularly advantageous on heavy gradients, and it was hoped that the scheme would reveal opportunities for considerable acceleration of service, in addition to economies in working. In Cornwall, for example, while it was likely that the overall speed limit of 60 mph would have to remain, if the heavy gradients could be climbed at 45–55 mph instead of 25–30 mph much faster overall times could be achieved. Unfortunately, the density of traffic all the year round and the prospects for its increase did not justify the high capital cost, and this ambitious project was abandoned, to the joy of many railway enthusiasts who had been aghast at the idea of the cherished Great Western steam locomotives being replaced by 'juice-boxes'.

The outbreak of the Second World War in September 1939 hit the British railways harder and more instantaneously than anything that had happened in 1914. In anticipation of widespread bombing attacks the government did everything possible to discourage all but the most essential of passenger travel. From Monday 25 September, with the exception of one night train, all services from London to the West of England ran via Bristol, and a substi-

36. Wartime austerity on the 'Limited' in 1943: coaches in plain brown, and the engine in plain green, devoid of any lining or polished work; with black-out curtain on the cab, and side windows sheeted over. The engine is no. 6022, 'King Edward III', and the train is passing near Reading West station. M. W. Earley

tute 10.30 a.m. from Paddington was not due at Plymouth till 4.40 p.m. instead of 2.35 p.m. as before the war. It carried the passengers formerly conveyed by the 10 a.m. 'Bristolian', by the 'Limited' itself, and by the 11.15 a.m. to Bath, Bristol and Weston-super-Mare. This ultra-austerity arrangement lasted only till 30 October, when running via Westbury by the 10.30 a.m. departure from Paddington was resumed, non-stop to Exeter, but at an average speed of only 48½ mph. Even this was a little faster than the recommended maximum laid down by the Ministry of War Transport, which was 45 mph.

At first speed was limited to 60 mph everywhere; but in 1941, with the virtual end of the night 'blitz', the limit was raised to 75 mph on many routes, including that between Paddington and Exeter. In the summer of 1941, with severe restrictions on the use of petrol for private motoring, vast numbers of people turned to the railways for their brief wartime vacations and on the Saturday before August Bank Holiday the 'Limited' ran in five sections with the following tremendous loading:

Section	Destination	Number of passengers
First	Penzance	743
Second	Penzance	1,089
Third	Paignton	1,000
Fourth	Exeter and Kingswear	800
Fifth	Newton Abbot	890

I may add that on the same day the 10.35 a.m. to Penzance running via Bristol carried in its two portions 1,276 passengers, though no doubt few of these would be travelling through from Paddington to Penzance. But there was little of the old-time holiday spirit about journeys on the 'Limited' in those years. It merely provided a brief escape from the toils of the war.

5

Post-war Austerity and Resurgence

By the early summer of 1946 the Great Western was making a brave attempt to restore passenger train services to something approaching their pre-war standards. The expresses to Cornwall were accelerated a little, and in Swindon works under the direction of F. W. Hawksworth the first steps were being taken to make the existing standard locomotive stock more capable of operating satisfactorily on the quality of coals that were then available. On the stationary testing plant I saw one of the Kings on trial after the fitting of a new form of superheater, a change that eventually proved the basis of a thorough rejuvenation of the class. But that first post-war summer was followed by the terrible winter of 1946–7, when appalling dislocation of traffic under snow and ice so impeded the delivery of coal that drastic reductions of train services had to be made and the 'Limited' itself was taken off. A substitute was run at 11 a.m. from Paddington, stopping additionally at Taunton, Exeter and Newton Abbot, but when the holiday season came round as many as four relief services to the 11 a.m. had to be run.

In the meantime there was a revival, in a totally different form, of the concept of making Cornwall a land apart so far as motive power was concerned. The coal shortage did not end with the return of normal weather in the late spring of 1947. I remember well the driver of a South Wales tank engine complaining to me that his engine would not steam on American coal – on the Taff Vale of all lines! I wonder what the management of that railway in its independent days would have thought of any coal, good or bad, being imported into South Wales. In consequence of such difficulties Hawksworth embarked on a limited programme of converting certain main-line locomotives to oil-firing, selecting engines of the Castle, Hall and the 2800 class of 2–8–0 for the purpose. While the first units to be converted were kept where they could conveniently be watched from Swindon, and were stationed at sheds like Bristol Bath Road, Old Oak Common and at Swindon itself, the ultimate intention was to make Cornwall an oil-fired area and so, as with the electrification project of pre-war years, avoid the cost of carrying locomotive coal to the West Country. Of course some coal supplies would have been needed at Plymouth and Newton Abbot to nourish the engines on the long distance runs to the east and on the

37. New power in the west: a Hawksworth 4–6–0 no. 1019, 'County of Glamorgan', on the Penzance–Wolverhampton through express at Dainton summit. The ganger holding up a dead rabbit causes great amusement to the engine crew as they pass. The engine standing on the down road is of LMS type – probably a Swindon-built '8F' 2–8–0.
John Ashman FRPS

west to north duties; but everything working in Cornwall would be oil-fired.

The plan was well under way by the summer of 1947, and was taken up with enthusiasm by the men of Laira shed. The Castle class engine no. 5079 'Lysander' was one of the first to be converted, and down in the West Country she came to be regarded as the flagship of the new fleet. For a time that summer, after the 'Limited' had been restored, she worked it both ways in Cornwall. This was considerably more than a run from Penzance to Plymouth and back; she made two return trips from Plymouth, with two separate Laira crews. Down in the early morning, with one of the night trains from Paddington; back with the up 'Limited'; westwards again with the down 'Limited'; and then back to Plymouth with a late evening stopping train – 320 miles in the day. I had the pleasure of riding on her footplate on the double trip from Penzance to Plymouth and back, and found her a splendid engine. The fireman had fully mastered the art of regulating the oil supply, and she steamed very freely. Indeed, one driver in his commendation of her performance said: 'You just can't knock her off the red line' – referring to the mark on the pressure gauge indicating maximum boiler pressure. I found 'Haberfield Hall' another very good oil-fired engine which

1006

39. Memories of Brunel in the West Country: a Hall and a County together take a heavy West of England express beneath the characteristic all-over roof of Exeter St Thomas station.

W. Philip Conolly

also took her turn on the 'Limited' in Cornwall.

The oil-firing project came to an end in one of those colossal muddles that all too frequently occur when officialdom intervenes in matters it does not understand. The success of the Great Western scheme reached the ears of the hard pressed Ministry of Fuel, and was hailed as a heaven-sent means of saving coal. Other British railways were urged to do the same. Vast sums of public money were spent in erecting oil-fuelling plants, before the whole thing came to a grinding halt: the Treasury could not find the necessary foreign exchange to buy the oil!

In the West of England the holiday traffic of 1948 was almost overwhelming. Not only did the habitual visitors to Cornwall flock to their old and favoured haunts, but the nationalisation of the railways which had taken place in January of that year brought something new. Railwaymen from far and wide, with travel facilities now extending over the whole country, determined to see for themselves what the Cornish Riviera had to offer. I well remember the stationmaster at the Kyle of Lochalsh telling me that he had booked up a holiday at Penzance! From the viewpoint of impressing visitors with what the Western

38. 'Counties' were frequently used on the 'Limited' in Cornwall after their introduction in 1945. Here, on 1 August 1951, no. 1006, 'County of Cornwall', is hauling a down West of England express beside the Exe estuary between Starcross and Dawlish Warren.

Norman E. Preedy

40. On the 1 in 37 eastbound approach to Dainton summit, the up 'Limited', King hauled, needs a Manor as bank engine, the latter going only as far east as Newton Abbot. The impressive façade of Dainton tunnel shows how much wider the bore needed to be to accommodate the broad gauge.

Mike Esau

could do in the running of trains, the best services, including the 'Limited' itself, were not available to pass and privilege ticket holders; but the crack trains like all the rest became engulfed in the tremendous weekend congestion that used to develop south-west of Taunton, and the holiday trains followed each other, block by block, all the way to Newton Abbot. Three other factors added to the congestion that prevailed on most Saturdays during the summer season. The introduction of holidays with pay meant a potentially bigger clientele anyway; petrol was still rationed, and comparatively few people could run cars; and the seaside landladies resolutely refused to take any bookings save from Saturday to Saturday.

A little time elapsed before the impact of nationalisation of the railways began to affect the Cornish Riviera. The locomotive interchange trials of 1948 were conducted on trains other than the 'Limited' and, although there were experiments with different styles of painting, these did not last very long. In his old age Sir Felix Pole was said to have remarked that in one way he was glad he was blind, and so spared the sight of a King painted blue. But nationalisation came to affect the men and equipment of the Great Western in a more insidious way. As this malaise was eventually fought and conquered in the running of the 'Cornish Riviera Express', it is worth dwelling upon it for a few lines. The top management of the GWR was bitterly opposed to nationalisa-

41. Post-nationalisation power on the West of England road, with a name reminiscent of the broad gauge. A Britannia class Pacific no. 70018, 'Flying Dutchman', on the 1.30 p.m. Paddington–Penzance express near Reading West Britannias were used on the 'Limited' in Cornwall, but not on the train east of Plymouth.

D. Hepburne-Scott

tion and, while the sentiments of very many of the staff had been favourable to it in principle, those feelings turned to bewilderment and frustration when after a few months they realised that many of the cherished traditions of the old Company were being swept away. In meeting their opposite numbers from other regions inborn resentment caused the Western to withdraw, involuntarily perhaps, from the comity of British Railways, and this in turn led to a degree of 'ganging up' from the others.

In locomotive matters, following the relatively disappointing results returned by the ex-GWR engines in the interchange trials of 1948, there was a general feeling that although the standards of workmanship at Swindon were impeccable, design practice had fallen somewhat behind the times. Little was known of the careful experimental work that was gradually bringing Kings and Castles up to the finest standards of the day. When, in the early 1950s, the General Manager of the Western Region, K. W. C. Grand, wanted to restore pre-war speeds with the best expresses, the operating department, echoing sentiments commonly held on the other regions of British Railways, argued that the Swindon locomotives were not capable of doing it with the quality of coal then available. Dynamometer car test runs with engine no. 6001, 'King Edward VII', in 1953 convinced the operators to the extent of restoring the 'Bristolian' to its

42. The up 'Limited' in Cornwall, passing Bodmin Road in September 1955 on the long climb up the valley from Lostwithiel and Doublebois. The engine, 4–6–0 no. 6913, 'Levens Hall', is in the lined-black livery of British Railways.
S. C. Nash

43. Another phase in post-nationalisation engine liveries: the engine, 4–6–0 no. 6027, 'King Richard I' is in the blue livery used for a time on the class 8 BR locomotives. This picture, taken in 1951 at Twyford, shows also the new style of painting on the carriage roof boards. M. W. Earley

44. Summer Saturday working on the up 'Limited' in 1959: a 14-coach train, double-headed by two Halls through from Penzance, is approaching Aller Junction. About a mile farther on it will have stopped at Newton Abbot west box for a King to take over for the non-stop run to Paddington. Derek Cross

$1\frac{3}{4}$-hour run in 1954, with King class engines and no more than a seven-coach load; but all the previous arguments were raised again when Grand wanted to restore the four-hour Paddington–Plymouth schedule of the 'Limited' for the summer service of 1955. Fortunately Alfred Smeddle, then Chief Mechanical Engineer at Swindon, was as forthright as Grand himself, and in as many words he said: 'If the Operating Department think the Pacific engines of other Regions can do better than our Kings, let us try them, and see!'

So a fascinating series of new interchange trials was projected entirely on the 'Cornish Riviera Express'. It did not prove quite so comprehensive as Smeddle originally intended due to the intransigence of one of the Eastern Region officers; but two other Pacifics were compared with the Kings – one theoretically, and the other by actual runs to and from Plymouth. At that time the Swindon locomotive testing section was putting the new British Railways standard three-cylinder Pacific, 'Duke of Gloucester', through its paces, both on the stationary plant and by dynamometer car tests on the line; and from the very comprehensive results obtained they had plotted the predicted performance of the engine with great accuracy over the West Coast Main Line between Euston and Carlisle.

As a collateral exercise they were able, with equal facility and accuracy, to plot its performance on the 'Limited' between Paddington and Plymouth. Then as a 'living' demonstration, to which the most senior operating men were invited, dynamometer car trials were made on the 'Limited' itself, made up to maximum summer load and run to the proposed four-hour schedule but in rough early spring weather. And in these conditions first a King and then a Duchess class Pacific of the London Midland Region were tested. From the outset the 'Duke of Gloucester' was ruled out of the comparison, because during the trials at Swindon its boiler could not equal the maximum output of one of Smeddle's rejuvenated Kings, but in the road trials the London Midland engine 'City of Bristol' did some fine work.

I was privileged to accompany two of the test runs, and of the many occasions on which I have travelled by the 'Limited' none remains more vividly in the memory than these two, experienced from the incomparable vantage point of the dynamometer car. The first came on a rough day in March when the trials were concerned with the Kings. In view of the results achieved, it is important to emphasise that the engine tested, 'King Henry VIII', was far from a picked unit. It was the first engine normally available and was if anything somewhat below average, with a badly leaking gland on the left-hand outside cylinder. With load made up to the full tonnage programmed for the summer service – fourteen to Westbury and twelve thence to Plymouth – the aim was to see how much time there was in hand with the engine worked at normal full capacity the whole way to Exeter. In the ordinary way, on a good unchecked run there would be opportunities to ease down a little in the later stages, particularly in the traditional place, over the Langport cut-off; but on this test run, if all was going well on the footplate, there was to be no let-up. A truly grand run was made. Despite two checks for permanent way repairs that cost 3½ minutes between them the Westbury slip portion was detached at Heywood Road Junction, 94½ miles, in a few seconds over 91 minutes, and the ensuing 79 miles to passing Exeter took only 68 minutes. The total time of 159¼ minutes from Paddington showed a gain of 8¼ minutes on the special test schedule.

A few weeks later the Stanier Pacific engine no. 46237 'City of Bristol' was on the job, but worked by regular Western Region men, who had been given plenty of opportunity beforehand to take the measure of the visiting engine. It was in this important respect that the competitive trials of 1955 differed from those of 1925 when the LNER Pacific engine ran to Plymouth and was worked by her own driver and fireman, guided by a road pilotman. In 1955 the 'City of Bristol' gave us an excellent run down, with a gain of 7 minutes to passing Exeter; but, as with 'King Henry VIII', a load of 420 tons carried forward after the Westbury slip had been detached was too great to be taken unaided over the exceptional gradients of the South Devon line, and a stop was made at Newton Abbot to attach a bank engine. These comparative trials, combined with those that had been obtained with the 'Duke of Gloucester', at last convinced the Operating Department of the Western Region not only that the Kings could do all that was required by the projected faster schedules, but that the big Pacific engines could do little, if any, better. And so for the summer service of 1955 the 'Limited' was once again booked to reach Plymouth North Road in the level four hours from Paddington.

On Saturdays in the height of the summer the 'Limited' was regularly booked in two sections, leaving Paddington at 10.30 and 10.35 a.m. The first section made its first passenger stop at Truro; the second section stopped first at Plymouth. Both were very heavy trains usually loaded to fourteen coaches, and both had to stop at Newton Abbot for bank engines. An interesting working was developed. Because the first part had no passenger stop in the Plymouth area it was arranged for the King that had worked down from London to couple off and double head the 10.35 forward to Plymouth, while two fresh 4–6–0s took on the 'Limited' itself to run the $85\frac{1}{2}$ miles to Truro non-stop. The Kings were not allowed to cross the Royal Albert Bridge at Saltash, and on those Saturdays the 'Limited' had various combinations of engines. I have seen two Halls, a Manor and one of the new Counties, while on a day when I had a footplate pass we had 'Carodoc Grange' and 'Wolseley Hall' to haul a fourteen-coach train in which practically every seat was taken – about 500 tons behind the second tender. The King that had brought us down from London had passed Exeter on time, but we were checked in the approach to Newton Abbot and got away on our non-stop run to Truro five minutes late. It was remarkable that on a

45. Momentous locomotive dynamometer car tests in 1955. The 'Limited' (14 coaches), hauled by 4–6–0 no. 6013, 'King Henry VII', approaching Heywood Road Junction, where the 2-coach Westbury slip portion was detached at full speed. The author was riding in the dynamometer car.

Kenneth Leech

46. Comparative tests in 1955 with an ex-LMS Duchess class Pacific: no. 46237, 'City of Bristol', again with a 14-coach train near Reading West. The author was in the dynamometer car on this occasion also.
M. W. Earley

day of very heavy traffic we got an absolutely clear road until arriving outside Truro, and for me it was a novel experience to ride through the Plymouth area on the footplate with nothing but clear signals ahead of us. For the record, we covered 83¾ miles, to a point on the outskirts of Truro, in 124¾ minutes from Newton Abbot, and although this represented a gain of nearly ten minutes on schedule it emphasises the difficulty of the line traversed by the 'Limited' west of Newton Abbot.

This Saturday spell from Paddington to Truro was actually not my longest on the footplate of the 'Limited', because in the autumn of 1955, when the faster schedules were in operation, I rode right through from Paddington to Penzance. By that time of year traffic had slackened off a good deal, and with the engine 'King Edward VIII' there was no difficulty in reaching Plymouth in the scheduled four hours; but there, to my interest, the engine waiting to take us forward to Penzance was the first of the Hawksworth County class 4–6–0s, and at that time the only one with a twin-orifice blastpipe and double chimney, no. 1000, 'County of Middlesex'. With a load of about 300 tons it was a relatively simple task for engine and crew, but my chief recollection of a pleasant occasion is of the scenic delights of the ride through Cornwall. It was a sunny, if rather boisterous, day in late autumn, and the countryside was looking its finest, with the trees in the valleys in their most gorgeous colouring and the open moors showing that astonishing clear air that only Cornwall seems to produce on a windy, sunny day. When we finally ran along the causeway from Marazion and came to rest in Penzance station, dead on time, I could have wished for more, even after $6\frac{1}{2}$ hours on the footplate.

By that time in history, however, the plan for complete modernisation of the British Railways had been launched, and one of its most important features was to be the complete elimination of steam traction. In high quarters it was naïvely thought that steam was at the root of the economic ills which were most seriously affecting British Railways at that time. By the mid-1950s the flood-tide of passenger business that had come after the end of the Second World War had receded to a degree far worse than anything in the slump of

47. The appearance of the County class 4–6–0s, so familiar on the 'Limited' west of Plymouth, was not improved by the curiously short double chimney put on after they had been re-draughted. Engine no. 1008, 'County of Cardigan', climbing the bank from Bodmin Road in 1960, shows this disfigurement prominently. Mike Esau

48. In 1953 when part of the Berks and Hants extension line was temporarily closed for extensive repairs to the road bed, the 'Limited' ran via Swindon. On weekdays, hauled by Castles, it took the Wilts and Somerset connecting line from Thingley Junction to Westbury, but on Sundays it ran via Bath and Bristol, King hauled. In this picture, hauled by 4–6–0 no. 6027, 'King Richard I', it is emerging from Middle Hill tunnel, Box. George Heiron

CORNISH RIVIERA
6027

49. The up 'Limited' at Hemerdon summit, South Devon, in 1959, just coming off the 1 in 42 gradient from Plympton, hauled by 4–6–0 no. 7812, 'Erlestoke Manor', and no. 7025, 'Sudeley Castle'. The latter, which will continue unassisted from Newton Abbot, is carrying the 'Cornish Riviera Express' headboard.

Derek Cross

50. The last phase of steam power on the 'Limited'. Engine no. 6022, 'King Edward III', with twin-orifice blastpipe and double chimney passing through Sonning cutting in 1958. M. W. Earley

the 1930s. The Cornish Riviera as a holiday region, so assiduously fostered by the Great Western Railway in former years, was as popular as ever, but visitors were travelling in their own cars, were towing caravans, or were bringing their own boats on trailers behind their cars. The traffic congestion that one could watch with such interest on the sea wall between Dawlish and Teignmouth had now passed to the highways, with cars queuing up for miles on the approaches to Exeter, Okehampton, Launceston and Bodmin. Yet some simple souls within the British Transport Commission persisted in their facile assumption that all would be well again on the railways once they had got rid of steam!

In the meantime the spirit of the old Great Western Railway lingered on, even though the Company had ceased to exist as such at the end of 1947. Ever since the moment that Brunel persuaded the original board to adopt the 7 ft gauge the GWR had a way of doing things differently from everyone else. In signalling, in traffic working, in its form of the vacuum automatic brake, there were outstanding points of difference, and in 1955, when everywhere else on British Railways thoughts were directed towards the large-scale introduction of diesel-electric traction, the Western decided on the hydraulic system of transmission. The policy of the Railway Board at that time was to try on a limited scale as many different types of diesel locomotive as possible with a view to making the best choice for future standardisation, and the Western project eventually provided much valuable experience. Naturally, the 'Limited' was the first train on which the new power was used, with many mixed feelings on the part of those who saw old and cherished traditions slipping away.

6
The Diesel Age

The introduction of the first diesel-hydraulic main-line locomotives in 1958 was not the first instance of non-steam traction on the GWR and the Western Region. The diesel railcars with mechanical transmission became quite an institution in the 1930s, and Hawksworth's experimental gas turbine locomotives came shortly after the war. But until 1958 nothing had occurred to disturb the regular and almost exclusive use of Kings on the 'Limited' between Paddington and Plymouth from 1927 onwards. In the disruption sometimes caused by war conditions other 4–6–0s may occasionally have been resorted to, but apart from that, and the use of London Midland Pacifics in the trials of 1955, the only regular – if temporary – use of other engines was in the autumn of 1951, when for a whole month the line was closed between Patney and Westbury and the 'Limited' was diverted via Swindon and Trowbridge. At that time of year the load leaving Paddington was usually less than 400 tons, and the train was worked by Castles. The train was allowed 214 minutes for the 187·8 miles to Exeter by that route, but there was usually plenty of time in hand.

Use of engines other than Kings on the 'Limited' has hitherto been a matter of extreme expediency since 1927, so it was highly significant of the changes in store when it was made one of the first jobs for the new diesel-hydraulic locomotives of the 600 class delivered from the North British Locomotive Company in 1958. The timing was then the level four hours, and I had a very interesting run down in the cab of the 'Ark Royal', the second locomotive of the class, later that year. Quite apart from the technique of driving the diesel, which was so different and apparently easy after the multifarious tasks to be done by the crew of a steam locomotive, there was the uninterrupted look-out ahead. It is astonishing to recall how constrained and one-sided the look-out was from the footplate of a big steam locomotive, especially those with Belpaire fireboxes, and until that first trip on a diesel I had never fully realised how sinuous the West of England main line is. The Berks and Hants section has very little straight track between Reading and Savernake, and the long sweeping curves of the cut-off lines are impressive in their fine engineering. Then, after the old main is rejoined at Taunton, one sees the curves south of Wellington with something akin to awe, in recalling that there, back in 1904, 'City of Truro' did the 'ton'!

As for the South Devon line, words failed me

51. The new age on the 'Limited': the Warship class diesel-hydraulic locomotive no. D 805, 'Benbow', photographed broadside-on while speeding westward through Sonning cutting in June 1959. British Rail

after that first ride on a diesel! Over most of it, west of Newton Abbot, speed was limited to 40 mph and in many places, as in going down to Totnes from Dainton tunnel, or in winding across the southern slopes of Dartmoor between Brent and Hemerdon, even this seemed a little too fast. I thought of the intrepid steam locomotive enginemen who used to take Saints and Stars over these stretches at about 60 mph, not to mention the hurricane flight in the reverse direction when 'City of Truro' did 77 mph east of Brent! With the diesels speed was always kept very strictly to the rules, and with good reason too, because they had not been long in service before it was realised that all was not well with their riding. After an impressive and much publicised debut the D 600 class was found unsatisfactory in many respects. The 800 series, the Warship class proper, were much better as locomotives, but one of them holds the unenviable distinction of being the only locomotive, anywhere in the world, on which I have been really scared. It was not ordinary rough riding, which anyone with a modicum of footplate experience is prepared to take as part of the job; it was a vicious sustained hunting, which at speeds of from 85 to 90 mph, was really disquieting. After

52. The 'Limited' diesel-hauled by the first of the North-British built D 600 class locomotives, no. D 600, 'Active', crossing the Royal Albert Bridge into Cornwall in April 1958. British Rail

53. Contrast at Penzance: the 'Limited' with the steam 4–6–0 no. 6949, 'Haberfield Hall', leaving in July 1953. C. R. L. Coles

54. A few years later the train leaving hauled by the North British built diesel-hydraulic locomotive no. D 603, 'Conquest'. John Ashman FRPS

two such experiences I sought no further rides on Warships for some time to come, and I was glad to learn that a maximum speed of 80 mph had been imposed upon them.

I avoided the larger Western diesels too, because in addition to being equally bad riders they were damnably hot in their cabs. It was later revealed that there were also problems with the transmission, vibration, excessive wear of parts and torsional oscillation, all pointing to the conclusion that the locomotives had been put into traffic before all the features of design had been fully proved. How the symptoms that gave rise to such alarmingly bad riding were studied, analysed and finally eliminated was a superb piece of mechanical engineering research at Swindon. The bogies of the Warship class were the same as those of the Krauss-Maffei V200 class locomotives in Germany; but, although these latter seemed to have acquired a good reputation there, they did not stand up to Western Region conditions. That most distinguished of latter-day British experimental engineers, S. O. Ell, summed it up in a memorable witticism:

> Preceptors of old advised us what happens to houses built on sand, and on rock, but be-

PENZANCE
D603
CORNISH RIVIERA
EXPRESS

cause the material had not then been invented, they could not advise us what happens to a house which is built on rubber stilts. This momentous discovery was left to British Railways. Hence the replacement of the rubber with a metal block and the end of the troubles with Krauss-Maffei bogies due to 'under design'.

In the meantime, while the engineers of the locomotive testing section were conducting their vital and ultimately successful work on the German type bogies, a new broom was beginning to sweep very clean in the headquarters offices at Paddington. Gerald Fiennes had been appointed General Manager, and fresh from the Eastern Region and the exploits of the Deltic locomotives he wanted substantial speeding-up of passenger train services. One of the first steps towards this objective was to run a seven-coach special from Paddington to Plymouth with a Western class locomotive, non-stop, in 208 minutes for the 225·5 miles. This spectacular feat, accomplished in April 1964, involved an overall average speed of 65 mph. But while a seven-coach train might be adequate for a 'hot-shot' businessman's flyer between London and Bristol, the call for

such a train to and from Plymouth was really minimal; and even with traffic declining as it was seven coaches would not do for the 'Cornish Riviera Express'. For the next trial a pair of Type 3 diesels were used with a ten-coach train. This provided a total of 3,500 horsepower, and a stop at Exeter was made to pick up some local newspapermen who, it was hoped, would be duly impressed.

The special was booked to reach Exeter, 173·5 miles in 136½ minutes. I was privileged to ride in the train, and with scrupulous observance of all speed restrictions, but very fast hillclimbing, we reached Exeter 4 minutes early at an average speed of 78·8 mph from Paddington. And we covered the remaining 52 miles on to Plymouth, with all their hindrances, in 62 minutes 6 seconds. The total time from Paddington, including the Exeter stop, was 196 minutes 17 seconds – an overall average speed of 68·8 mph. Unfortunately Fiennes did not remain at Paddington long enough for his ambitious programme to be implemented, and the overall time of the 'Limited' between Paddington and Plymouth remained at the level four hours for some years after these exciting trial runs. The insertion of extra stops within the same overall timing certainly made necessary some faster point to point running over intermediate stages. By 1967, for example, the 'Limited' made its first stop at Taunton, in 123 minutes for the 142·7 miles, an average of 69·4 mph. On this timing the 2,700-horsepower Western class diesel-hydraulic locomotives took loads of up to 500 tons at busy times, but for the most part the business was getting progressively lighter. Spectacularly faster services such as those envisaged by Gerald Fiennes were not considered likely to win back traffic.

The Cornish Riviera service was conceived as a holiday traffic, and a holiday traffic it remained for good or ill. But by the 1960s the whole situation was changing. Cornwall still attracted its tens of thousands, but they did not come by train. They came in their own cars. Analyses of business on the branch lines laid them wide open to the Beeching axe. There was a time indeed when there was a threat to close the entire railway west of Plymouth! Happily such drastic thoughts were put on to one side, but today the 'Cornish Riviera Express' and the railway in Cornwall as a whole lie at the very heart of a modern problem in transportation – a problem that is as much sociological as it is one of railways – and it is interesting to speculate on what the eventual outcome will be.

The time has long passed when the family holiday began at the moment of departure from home, when the journey was an adventure and an exciting prelude to a stay at the seaside. Now, even with long stretches of motorway on which one can legally drive at 70 mph for hours, the holiday peaks bring frustrating hold-ups and exasperated and venturesome driving, and journey's end can be reached in a state of nervous exhaustion. One feels that many travellers would prefer the quiet, effortless journey down to Cornwall by train but would need, above all, the convenience of having their car on hand at their chosen holiday place. Again, however, the congestion around the most favoured resorts in the height of the

55. Another contrast: the down 'Limited' hauled by one of the D 800 class Warships passing alongside the Southern Region Meldon quarry ballast train in Exeter St Davids station. W. Philip Conolly

56. The diesels need assisting too on the South Devon line. A Warship on the up 'Limited' approaching Aller Junction double-headed by a Castle – no. 5003, 'Lulworth Castle', and D 806, 'Cambrian'. Derek Cross

season is becoming intolerable. The fact that it is accepted as 'just one of those things' does not alter its acuteness. One wonders if the enterprising management of the Great Western Railway, which eighty years ago started boosting the Cornish Riviera as a holiday playground for all the year round, ever foresaw such a situation as now exists.

Yet in its possession of a finely maintained double-track main line extending to within ten miles of Land's End, British Railways have the roadbed of the finest tool for bulk transportation, and potentially the most economical, yet devised by man. Today it is terribly under-utilised, while the highways are at times choked almost to a standstill. There are, of course, the motorail services, whereby private cars are conveyed by the same train as their owners, but the inheritors of the great spirit of enterprise that virtually created the Cornish

57. Diesel double-headed westbound: the 'Limited' ascending Dainton bank in August 1960 with 4–6–0 no. 6959, 'Peatling Hall', assisting no. D 812, 'Royal Naval Reserve 1859–1959', on a 12-coach train. Hugh Ballantyne

1B25

Riviera have an unrivalled chance, at this time of virtual crisis in private motoring, to come out with some imaginative and attractive package deal that would get the new minis and the old bangers onto a whole fleet of motorail versions of the 'Cornish Riviera Express', and land their patrons within easy reach of their chosen holiday retreats.

By their process of retrenchment, however, British Railways have made things rather difficult for themselves in any scheme of re-expansion. Some main-line stations have been closed, and while it would seem a rather striking change from former days that the 'Limited' of today calls at Bodmin Road, St Austell and Redruth, all of which it used to pass, one is somewhat startled to find that Bodmin Road is now the railhead for Padstow, because of the closing of the one-time Southern line in Cornwall, and the consequent elimination of the through express service from Waterloo. The 'Limited' certainly makes fast time to Plymouth, reaching there in 3 hours 34 minutes from Paddington, an overall average of 63·3 mph. But at Plymouth it sheds its restaurant car, and any passengers who have journeyed through from Paddington could well feel their tongues hanging out by the time Penzance is reached at 17·05. The overall time today is nevertheless 55 minutes quicker than in the fastest steam days.

The 'Limited' now leaves Paddington at 11.30, instead of 10.30, and it looks much the same as any other locomotive-hauled train of today on British Railways. One has to look at the rather inconspicuous paper 'stickers' on the end doors of the carriages to see that it *is* the 'Cornish Riviera'. It is of course entirely in keeping with the traffic policy of British Railways generally that there should be other trains on the Cornish service that are nearly as fast. The 9.30 out of Paddington, for example, reaches Penzance at 15.20, and the 13.30 at 19.19. The only additional stop made by these two trains is at Reading. In earlier days, however, the 12 noon and 3.30 departures from Paddington equalled the running times of the 'Limited'.

Shorn though it is of most of its old time glamour, it would be graceless to leave this famous train on a sour note, even though it slips almost incognito out of Paddington at 11.30. For those who are in a mood to appreciate it the train provides one of the most delightful railway rides to be enjoyed anywhere, now performed at considerably higher speed than at any time in the great days of steam. To be thoroughly nostalgic, a well-thumbed and much cherished copy of *Through the Window from Paddington to Penzance* should be one's companion, not forgetting of course that nowadays lunch is going to cost a good deal more than three shillings and sixpence! By the time this book is published the last of the Western diesels will have been withdrawn, and one's locomotive, in 1978, will probably be a Class 50 diesel-electric.

How partisanship can rage among railway enthusiasts, even over diesels! There were mixed feelings when the first of the diesel-hydraulics arrived to take the place of Castles and Kings, but they became accepted as the Western Region diesels, maintaining the GWR tradition of old by being different from all others. That both Warships and Westerns were a packet of trouble in their early days

58. A Western, no. D 1001, 'Western Pathfinder', climbing the sinuous track to Dainton summit in August 1971, amid the beautiful scenery of South Devon.
J. H. Cooper-Smith

was vaguely known, but not to the extent of technical expertise. But when they were cured and settled down to their allotted work they became as cherished units as ever the steam locomotives had been by an earlier generation. Then when British Railways, after the electrification to Glasgow was completed, dared to foist upon the Western Region the cast-off Class 50s from the West Coast main line the heather was ablaze. The fact that the Class 50s were not in the best of condition when they arrived, and that they had a great deal of rather sophisticated equipment into the bargain did not help to warm up the distinctly chilly reception they received on the Western Region; but the fans were beside themselves with indignation. That the Western Region enginemen have learned to do great things with the 50s is however a matter of history.

With a train like the 'Cornish Riviera Limited', as it is now known, there are inevitably many personal associations, from holiday trips, and in my own case many runs on the footplate of both steam and diesel locomotives. So perhaps I may end on a personal note. In the early summer of 1975 my wife and I enjoyed a holiday in the Cornish Riviera. Proofs of an urgent literary assignment had been late in coming and I had to give the publishers my holiday address. We arrived back to lunch one morning to find a telephone message: 'Proofs consigned to you by Red Star on 11.30 from Paddington. Please collect at Redruth.' 'Red Star' on the Limited! I am afraid my mind

59. Journey's end: a 47 class diesel, no. 47086, 'Colossus', bringing its train into Penzance on 7 July 1976.
Brian Morrison

60. Industrial archaeology, old and new, in Cornwall: the up 'Limited' in July 1977, hauled by a 50 class diesel electric locomotive no. 50004, passing the ruins of an old tin mine near Scorrier, near Redruth.

Brian Morrison

went racing back far beyond a bundle of overdue proofs, because the first time I saw the 'Limited' after I had begun my engineering studies in London in 1921, the engine was no. 4006 'Red Star'. She was then at Laira, and one of the regular runners on the Plymouth–London double home turns. The coincidence on this 1975 occasion was too striking to be passed over lightly, and that afternoon I fell to day-dreaming.

The hotel grounds led down to one of those exquisitely secluded tidal creeks that one can still find in Cornwall, away from all the bustle of the traditional holiday activities of today; and I sat for half an hour or so thinking of old times on the 'Limited'. It is, I am told, symptomatic of advancing years that one remembers clearly people and things of one's youth far more accurately than more recent happenings, when urgent tasks have to be dealt with promptly, finished, and attention concentrated on the next; and that afternoon I recalled the engines I had seen on the 'Limited' around London, at Reading, and down in the West Country: 'Queen Adelaide', such a favourite in 1925; 'King Edward', 'Malvern Abbey', 'Polar Star', 'Knight of the Thistle' – all in days before the Castles came upon the scene. Great days and great engines. Kipling wrote about romance bringing up the 9.15, but outwardly at any rate there was surely no train, not even the 'Flying Scotsman' or the even older established 'Irish Mail' to which the word romance could be more aptly applied than to the 'Limited'.

Later that afternoon we drove to Redruth to collect that parcel, and when the 'Limited' of today rolled in one's thoughts could well have been very far from romance – a short rake of standard coaches hauled by a diesel. And while I can still unfailingly remember the names of engines I saw on the train more than fifty years ago, I cannot now remember even what class of locomotive it was that brought the 'Limited' into Redruth that early summer afternoon only a couple of years ago. Was it a 50 or a Western? Yet this is unkind, and unworthy of me. When he wrote those well remembered lines, Kipling was not thinking of great international express trains, or what we should now call the Inter-City services. He was writing of the humble commuter train. The full quotation is:

'Romance!' the season tickets mourn,
'*He* never ran to catch his train,
But passed with coach and guard and horn –
And left the local – late again!'
Confound Romance! . . . And all unseen
Romance brought up the nine fifteen.

And today, as truly as when Kipling wrote those lines, romance is at the very heart of all railway operation, however drab the outward trappings may be. The 'Limited' lives on; and one day when our highways are so choked or petrol has become so expensive that private motoring is prohibitive, the engineering science that elsewhere in our country enables 100 mph express trains to run at five-minute intervals in perfect safety may assist in making that packaged deal for the mini-owners on the line to the Cornish Riviera.

61. On the sea wall at Teignmouth: one of the 50 class diesels emerges on to the sea wall – one of the most popular viewing points of West of England expresses even from broad-gauge days. This photograph was taken in March 1977 from the ladder of the Teignmouth down outer home signal.
John Goss

Index